The China Geological Survey Series

This Open Access book series systematically presents the outcomes and achievements of regional geological surveys, mineral geological surveys, hydrogeological and other types of geological surveys conducted in various regions of China. The goal of the series is to provide researchers and professional geologists with a substantial knowledge base before they commence investigations in a particular area of China. Accordingly, it includes a wealth of information on maps and cross-sections, past and current models, geophysical investigations, geochemical datasets, economic geology, geotourism (Geoparks), and geo-environmental/ecological concerns.

More information about this series at http://www.springer.com/series/16470

Fanyu Qi · Xiaolei Li · Yuntao Shang ·
Jie Meng · Xuezheng Gao · Zhaoyu Kong ·
Haixin Li · Haifei Yan

Atlas of Mineral Deposits Distribution in China (2020)

Fanyu Qi
Development and Research Center
China Geological Survey
Beijing, China

Xiaolei Li
Development and Research Center
China Geological Survey
Beijing, China

Yuntao Shang
Development and Research Center
China Geological Survey
Beijing, China

Jie Meng
Development and Research Center
China Geological Survey
Beijing, China

Xuezheng Gao
Development and Research Center
China Geological Survey
Beijing, China

Zhaoyu Kong
Development and Research Center
China Geological Survey
Beijing, China

Haixin Li
Foshan Geological Bureau of Guangdong
Province
Foshan, Guangdong, China

Haifei Yan
Geological Information Center ECE
Nanjing, Jiangsu, China

ISSN 2662-4923   ISSN 2662-4931   (electronic)
The China Geological Survey Series
ISBN 978-981-16-0974-9   ISBN 978-981-16-0972-5   (eBook)
https://doi.org/10.1007/978-981-16-0972-5

This Springer imprint is published by the registered company Springer Nature Singapore Pte Ltd.
The registered company address is: 152 Beach Road, #21-01/04 Gateway East, Singapore 189721, Singapore

Foreword

Mineral resources are of great strategic significance since they matter in every aspect of social and economic development. However, existing mineral resources are unlikely to support sustainable economic development given the planning and construction of urbanization in China. Therefore, it is a top priority to explore more mineral resources to sustain the steady development of the national economy, and sufficient data serve as the key to mineral resource survey.

In 2018, the Development and Research Center of China Geological Survey (also known as the National Geological Archives of China) started the pilot building of a new edition of the National Mineral Deposit Database of China. After more than two years' efforts, 307,081 pieces of mineral deposit data have been converged from the data holdings of the National Geological Archives of China. In this manner, the National Mineral Deposit Database of China (2020) was formed. While still being built, the database has served the needs of economic and social development and achieved significant effects.

The *Atlas of Mineral Deposits Distribution in China (2020)* (also referred to as the *Atlas 2020*) was prepared based on the data in the National Mineral Deposit Database of China (2020). It mainly reflects the distribution of metallic, non-metallic, and energy mineral deposits in 30 provinces, municipalities, and autonomous regions (Data are not yet available in Shanghai Shi, Hong Kong Sar, Macau Sar, and Taiwan Province) in China, thus assuming great significance for prospecting of mineral resources.

The *Atlas 2020* is the result of deep data mining, development, and utilization of geologic data. It was published to provide important basic data for the mineral geological survey, metallogenic study, planning of national land space, and construction of major projects in China. Therefore, it is expected to achieve substantial social and economic benefits.

We gratefully acknowledge: (i) relevant leaders for their strong support towards the publishment of the *Atlas 2020*; (ii) many experts, scholars, and peers for their valuable guidance and suggestions. In addition, concerning any omission or error in the *Atlas 2020*, the readers are encouraged to notify us for correction.

Beijing, China   Fanyu Qi

Contents

1.1 Background

The National Mineral Deposit Database of China (2002) (also referred to as the Database 2002) has been ranking top in terms of daily data distribution volume in the National Geological Archives of China (also referred to as the NGAC). However, it cannot be updated in a sustainable manner and cannot provide up-to-date, comprehensive data at present. Therefore, a new edition of the database is urgently needed by the industry of mineral geological survey and social users. To this end, the NGAC is committed to vigorously promoting the development and utilization of mineral-related geologic data holdings and comprehensively integrating mine-related geologic data. In addition, it has also made great efforts to study the update and maintenance mechanisms of mineral deposit data and produce thematic mineral deposit data to serve the needs of all parties. The purpose is to provide support for mineral geological survey, prospecting prediction, and research on metallogenic rules. Meanwhile, it has strived to comprehensively improve the informatization of geological survey and information service level through deep data mining of mineral-related geologic data. The aims are to meet the growing demand of mining industry development for mineral-related geologic data, to provide rich and reliable geologic data for economic and social development, to provide scientific and reliable research and assessment results for government decision-making, and to provide the most comprehensive mineral resource information for mining service fields.

1.2 Method

Geologic data contain a great deal of geological knowledge, which especially includes the information of mineral deposits. According to statistics of the data holdings of the NGAC, mineral exploration data account for more than 60% of all data holdings. In addition, the regional mineral survey data and geological scientific research data in regional survey data also contain a large amount of mineral deposit information. All these provide data sources for the building of a new edition of the National Mineral Deposit Database of China.

A new edition of the database is a comprehensive professional database established based on the Database 2002. The establishment process is as follows. The mineral deposit data are first mined from the mineral-related data holdings of the NGAC using the big data mining technology. Afterwards, the data are cleaned, sorted, integrated, extracted, converged, and input into the database (Fig. 1.1). It can be continuously updated and thus can dynamically reflect the up-to-date state of the mineral resources in China.

To serve the needs of social and economic development while still being built, a new edition of the database will be established once every year.

© The Author(s) 2021
F. Qi et al., *Atlas of Mineral Deposits Distribution in China (2020)*, The China Geological Survey Series,
https://doi.org/10.1007/978-981-16-0972-5_1

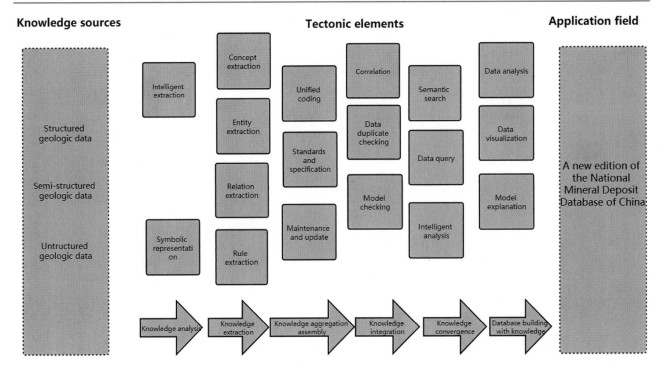

Fig. 1.1 Method of building a new edition of the National Mineral Deposit Database of China

Achievements Made in the Building of the National Mineral Deposit Database of China (2020)

2

2.1 Achievements in Data Convergence

By the end of June 2020, the Database 2020 contained 307,081 pieces of valid mineral deposit data, with an increase of more than 270,000 pieces compared to the 33,794 pieces in the Database 2002. In addition, the newly added content in the Database 2020 also includes the data tables such as code comparison table and the data items such as the code of 1:50,000-scaled map sheet of the mineral deposit data, category of major mineral types, and data source. Presently, the Database 2020 is the largest, most complete, and most comprehensive mineral deposit database in China.

In terms of mineral types, the total 307,081 pieces of data in the Database 2020 include 44,038 pieces of data on energy mineral deposits, accounting for 14.341%; 103,571 pieces of data on metal mineral deposits, accounting for 33.728%; 841 pieces of data on groundwater and gas mineral deposits, accounting for 0.274%; 158,631 pieces of data on non-metallic mineral deposits, accounting for 51.658% (Fig. 2.1).

The 103,571 pieces of data on metal mineral deposits include 37,339 pieces of data on ferrous metal deposits, accounting for 36.05%; 39,688 pieces of data on nonferrous metal deposits, accounting for 38.32%; 22,661 pieces of data on precious metal deposits, accounting for 21.88%; 1885 pieces of data on rare metal deposits, accounting for 1.82%; 555 pieces of data on dispersed element deposits, accounting for 0.54%; 1443 pieces of data on rare earth element metal deposits, accounting for 1.39% (Fig. 2.2).

2.2 Achievements in Data Mining

So far, the Database 2020 has integrated the data from the Database 2002, the National Mineral Deposit Database of China (by sectors), the achievements obtained from the Nationwide Mineral Resource Potential Assessment, the results of the Nationwide Mineral Resource Utilization Status Survey, the nationwide mineral reserves database of China, and the result reports, attached forms, and attachments involving mineral deposits in the geologic data holdings of the NGAC.

According to data sources recorded, in all the 307,081 pieces of mineral deposit data in the Database 2020, 217,151 pieces were extracted from the geologic data holdings of the NGAC using data mining technology, accounting for 71%. These data cover the results of a situation analysis and distribution of the mineral resources in China achieved by previous researchers over the past 100 years. In detail, they include national mine environment survey results, annals of minerals, mineral reserves sheet per year, and the data

© The Author(s) 2021
F. Qi et al., *Atlas of Mineral Deposits Distribution in China (2020)*, The China Geological Survey Series,
https://doi.org/10.1007/978-981-16-0972-5_2

Fig. 2.1 Statistics of data in the National Mineral Deposit Database of China (2020) (by mineral types)

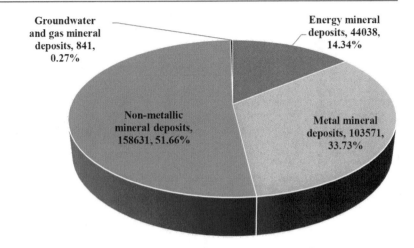

Fig. 2.2 Statistics of data on metal mineral deposits in the National Mineral Deposit Database of China (2020)

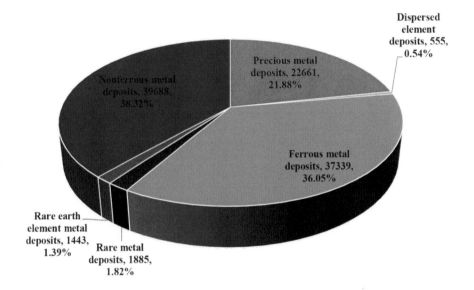

obtained during Manchukuo period (1932–1945) including mineral exploration data and mineral deposit cards in mineral geological survey results. Therefore, data mining was thoroughly conducted for all the mineral deposit-related geologic data holdings of the NGAC.

Significance of the Building of National Mineral Deposit Database of China (2020)

3.1 Providing Data for Smart Mineral Geological Survey of the China Geological Survey

Data are fundamental strategic resources of a country and are considered as "diamond mines" in the twenty-first century. Xi Jinping, general secretary of the Communist Party of China (CPC) Central Committee, has pointed out that big data is the next stage of informatization development and has significant and profound impacts on the economic development, social progress, and people's lives worldwide. He mentioned that all countries in the world regarded the digital economy as an important driving force to achieve innovation and development. He also stressed that China has paid close attention to the development of the digital economy, pushed forward the digital economy to promote high-quality economic development, implemented the national big data strategy, and accelerated the construction of Digital China. The State Council of China issued the *Action Framework for Promoting the Development of Big Data* in 2015, marking that the big data for development officially became a national strategy of China. It also indicates that big data is growing into a new powerful engine driving economic transformation and development, a new opportunity to reshape the competitive advantages of the country, and a new point of economic growth that significantly affects the future information industry pattern. In terms of natural resource management, geologic data are national fundamental strategic resources and play an important role in promoting the transformation and development of the national economy.

Currently, the China Geological Survey is developing the *Implementation Plan of Smart Mineral Geological Survey*. Data is particularly important in smart prospecting prediction. China has carried out four rounds of mineral resource survey so far, and has achieved a series of result data. Furthermore, a large volume of data will be continually generated every year through the mineral geological survey. These data will serve as the base of the smart mineral survey. Only with the support of abundant data can enough sample data be available for the smart geological survey and prospecting. The Database 2020 will converge the achievements of all kinds of mineral resources in China, and thus lay an important database for smart mineral geological survey.

3.2 More Efficiently Managing Mineral Resource Data of China

The Database 2020 is intended to serve the needs of natural resource authorities, management departments of geological exploration and mineral development, mining enterprises, and scientific research institutes. With it, the large volume of complex mineral resource data obtained from diverse data sources can be efficiently managed, thus facilitating the understanding of the distribution, scale, and potential of mineral resources in China. Meanwhile, using this database, the mineral resource data of China can be converged, managed, developed, utilized, and put into service, and more convenient functions such as query, retrieval, and download can be provided. In this way, all kinds of result data of mineral resource survey are comprehensively managed, providing more strong data support and higher-level services for users.

3.3 Improving the Up-to-Dateness of the Mineral Resource Data of China

The Database 2002 had long been one of the databases serving the users the most frequently in the NGAC, thus holding great service value. However, since no necessary update and maintenance mechanisms were established when

© The Author(s) 2021
F. Qi et al., *Atlas of Mineral Deposits Distribution in China (2020)*, The China Geological Survey Series,
https://doi.org/10.1007/978-981-16-0972-5_3

it was built, the mineral deposit data in the database has not been comprehensively and systematically updated for many years. This has seriously reduced the up-to-dateness of professional data.

In recent years, various mineral deposit databases have been submitted to the NGAC and the social demand for these databases has been gradually increased. Thus it was put on the agenda to build a more practical, more advanced mineral deposit data service system that can be continuously updated and shared based on the Database 2002. As various mineral resource databases were submitted to the NGAC, the original data in the Database 2002 can be updated by integrating the mineral data in these databases, thus improving the data up-to-dateness. So far, the mineral-related databases in the holdings of the NGAC include the result data of multiple national geological survey projects, such as the "Nationwide Mineral Resource Potential Assessment", the "Nationwide Mineral Resource Utilization State Survey", the "Superseding Resources Exploration in Resources Crisis Mines", and "Mineral Geological Survey Results". They provide the latest data and supplementary data for the building of a new edition of the *National Mineral Deposit Database of China*. Meanwhile, the Database 2020 will be updated based on the results of various mineral resource survey projects and integrate the data of all kinds of mineral deposits. Therefore, the up-to-dateness of the Database 2020 will be improved.

3.4 Providing Comprehensive Data for Prospecting Prediction, Scientific Research, and Mineral Geological Survey

In 2019, the NGAC organized over 20 mining enterprises, research institutes, and building material departments to propose and communicate their data demand face-to-face. This multi-industry demand survey provided specific and practical demand orientation for the NGAC to build the Database 2020. The project team members of the Database 2020 deeply understood the new changes in the demand for mineral resource data of prospecting professionals and scientific researchers, which laid a foundation for providing precise data services of mineral resources.

The Database 2020 was built with a view to better aggregate and integrate the thematic services of various mineral resource data and provide more precise service

models. The purpose is to provide precise, high-level, comprehensive mineral resource data services for different fields and users involving prospecting prediction, scientific research, and mineral geological survey.

3.5 A Typical Case of Building a Database Using Big Data Mining Technology Based on Geologic Data Holdings of NGAC

Geologic data are also the collection of all kinds of geoscientific data and contain a large amount of geological knowledge, such as a large volume of mineral deposit data. To extract geological knowledge from the massive geologic data resources by technical means and to integrate them into a database will not only significantly lower the building cost of a geologic database but also make it possible to efficiently update and maintain the data in the database. This is the core innovative concept in the building of the Database 2020.

Mineral deposit information is one of the geological knowledge points contained in geologic data. According to the statistics of geologic data holdings of the NGAC, mineral exploration data account for more than 60% of all the data holdings, exhibiting the important position of mineral exploration in previous geological work in China. At the beginning of the twenty-first century, the China Geological Survey organized its geological survey branches in various provinces to build a mineral deposits database. As a result, the Database 2002 was built at a high cost. It was an important basic geologic database of China and effectively preserved the geological and mineral exploration data of China. However, with the advancement of society, science, and technology, geological work has been constantly developing and changing. As a result, the data items in the Database 2002 can hardly meet the need of social services, and it is necessary to further improve the database model of the Database 2002. Furthermore, no data update mechanism was established when the database was built. This makes it difficult to timely update the database and to guarantee the up-to-dateness of the mineral deposit data in the database after more than ten years of geological work. The Database 2020 has been built by extracting geological knowledge from the mineral deposit data contained in geologic data using the big data mining technology and then converging the knowledge into a database. Thus the building of the

Database 2020 serves as a successful case of the concept of building a database by obtaining geological knowledge using big data mining technology.

3.6 Providing Convenient Data Query and Retrieval Services for Users with Online Service Platform

In the process of building a new edition of the National Mineral Deposit Database of China, phased results are obtained each year. So far, the Database 2020 has been built to provide social services. Furthermore, the online service platform of the Database 2020 has been recently developed and put into trial operation. The Database 2020 allows users to conveniently query mineral deposit data and filter the data by geographical division and metallogenic belts.

With the Database 2020, users can conduct online queries about all the basic information of mineral deposits stored in the database, including the name, longitude, latitude, mineral types, genetic type, scale, geological work degree, and geographical location. Users can query the mineral deposit information by the name and mineral type of mineral deposits and filter the information by administrative areas and user-defined conditions.

Fig. 3.1 Website of service platform app of the National Mineral Deposit Database of China (2020)

Website of the online service platform of Database 2020: https://data.ngac.org.cn/mineralresource/ (Fig. 3.1).

Maps of Mineral Deposits Distribution by Provinces in China

4

F. Qi et al., *Atlas of Mineral Deposits Distribution in China (2020)*, The China Geological Survey Series, https://doi.org/10.1007/978-981-16-0972-5_4

Mineral legend

Mineral type	Super-large	Ultra-large	Large	Middle	Small	Ore occurrence	Mineralized point
Andesite							
Attapulgite clay							
Cretaceous							
Dolomite							
Slate							
Gem							
Iceland spar							
Trachyte							
Marble							
Iodine							
Tourmaline							
Witherite							
Calcite							
Zeolite							
Pink quartz							
Peridotite							
Adamantine spar							
Kaoline							
Wollastonite							
Diatomite							
Sepiolite clay							
K-bearing sandshale							
K-bearing rock							
Obsidian							
Andalusite							
Granite							
Talc							
Gneiss							
Other clay							
Sandstone							
Diorite							
Serpentinite							
Arsenic							
Gypsum							
Limestone							
Garnet							
Asbestos							
Graphite							
Quartzite							
Quartz crystal							
Pitchstone							
Ceramic clay							
Trona							
Natural quartz sand							
Natural whetstone							

Mineral type	Super-large	Ultra-large	Large	Middle	Small	Ore occurrence	Mineralized point
Fluorite							
Jade							
Mica							
Feldspar							
Native sulphur							
Topaz							
Diabase							
Volcanic ash							
Cinder							
Potassium salt							
Amphibolite							
Diamond							
Mineral salt							
Disthene							
Blue asbestos							
Rectorite clay							
Phosphate							
Magnesite							
Pyrite							
Agate							
Granodiorite							
Vein quartz							
Mirabilite							
Magnesium sulfate							
Alunite							
Nitratine							
Refractory clay							
Marl							
Mudstone							
Tuff							
Boron							
Bentonite							
Laterite							
Diopside							
Tremolite							
Nepheline syenite							
Bromine							
Basalt							
Pigment mineral							
Pyrophyllite							
Shale							
Illite clay							
Perlite							
Vermiculite							
Barite							

Mineral type	Super-large	Ultra-large	Large	Middle	Small	Ore occurrence	Mineralized point
Vanadium							
Chromium							
Manganese							
Titanium							
Iron							
Tellurium							
Cadmium							
Hafnium							
Gallium							
Rhenium							
Thallium							
Selenium							
Indium							
Germanium							
Bismuth							
Mercury							
Cobalt							
Alumina							
Magnesium							
Molybdenum							
Nickel							
Lead							
Stibium							
Copper							
Tungsten							
Stannum							
Zinc							
Platinum							
Osmium							
Gold							
Ruthenium							
Palladium							
Iridium							
Silver							
Zirconium							
Lithium							
Niobium							
Beryllium							
Rubidium							
Cesium							
Strontium							
Tantalum							
Coal							
Stone coal							
Natural asphalt							

Geographical legend

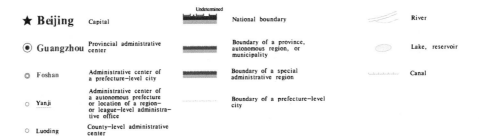

★ **Beijing**   Capital

◉ **Guangzhou**   Provincial administrative center

◎ Foshan   Administrative center of a prefecture–level city

○ Yanji   Administrative center of a autonomous prefecture or location of a region– or league–level administrative office

○ Luoding   County–level administrative center

Undetermined
▬▬▬ National boundary

▬▬▬ Boundary of a province, autonomous region, or municipality

▬▬▬ Boundary of a special administrative region

········· Boundary of a prefecture–level city

⌇⌇⌇ River

⬭ Lake, reservoir

·–·–·– Canal

Deposit distribution map of Beijing Shi, China

Deposit distribution map of Tianjin Shi, China

0   5   10 15 km

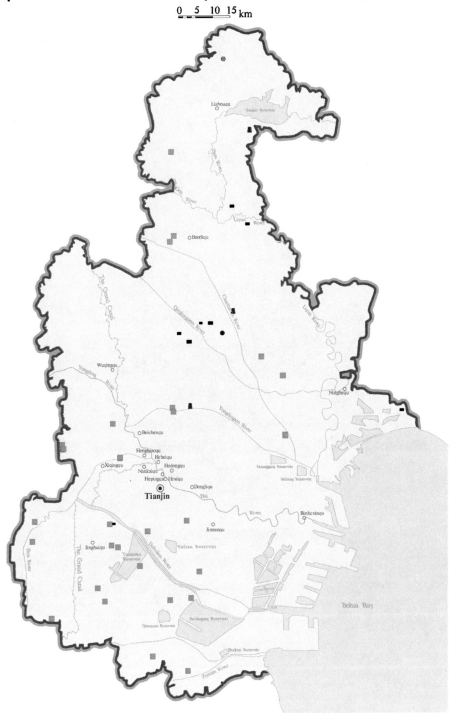

Deposit distribution map of Hebei Province
(metallic minerals)

0   20 40 60 km

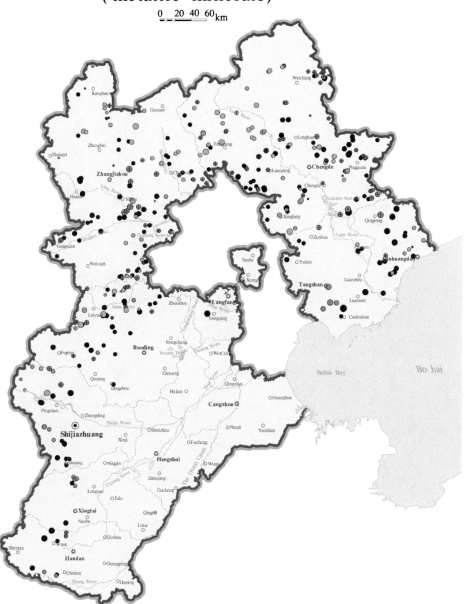

Deposit distribution map of Hebei Province
(energy minerals)

0 20 40 60 km

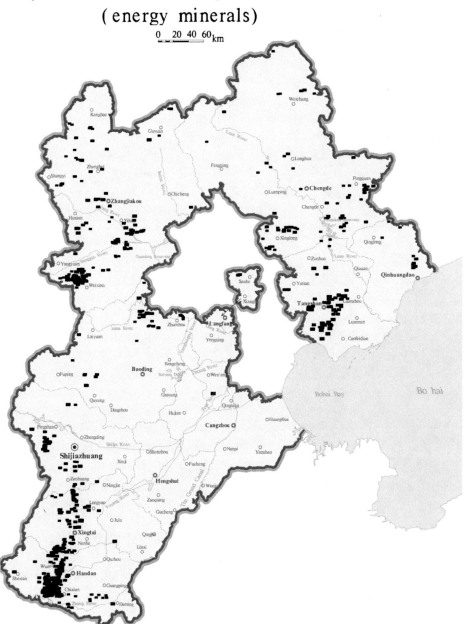

Deposit distribution map of Hebei Province
(non-metallic minerals)

0 20 40 60 km

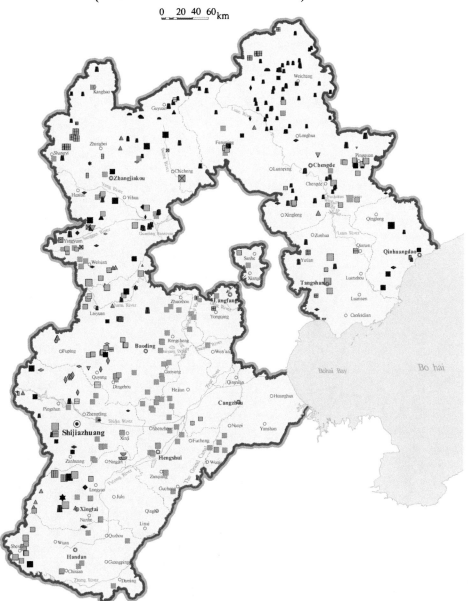

Deposit distribution map of Shanxi Province
(metallic minerals)

0   20   40   60 km

Deposit distribution map of Shanxi Province
(energy minerals)

0   20   40   60 km

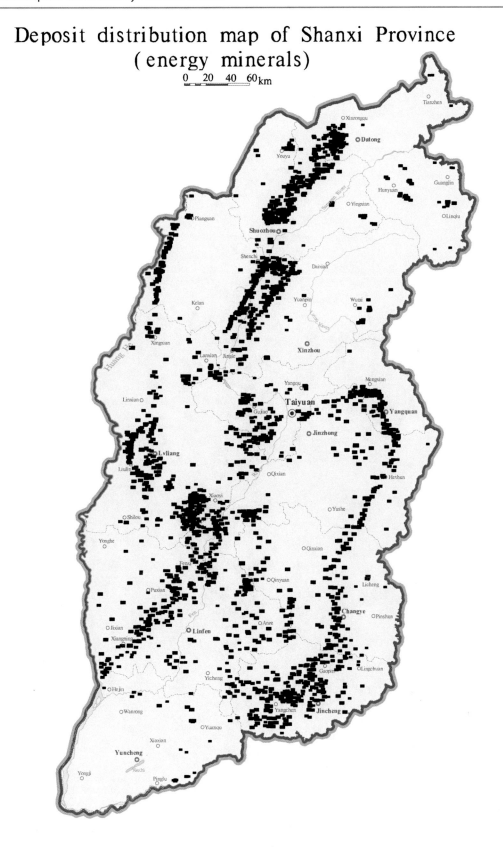

Deposit distribution map of Shanxi Province
(non-metallic minerals)

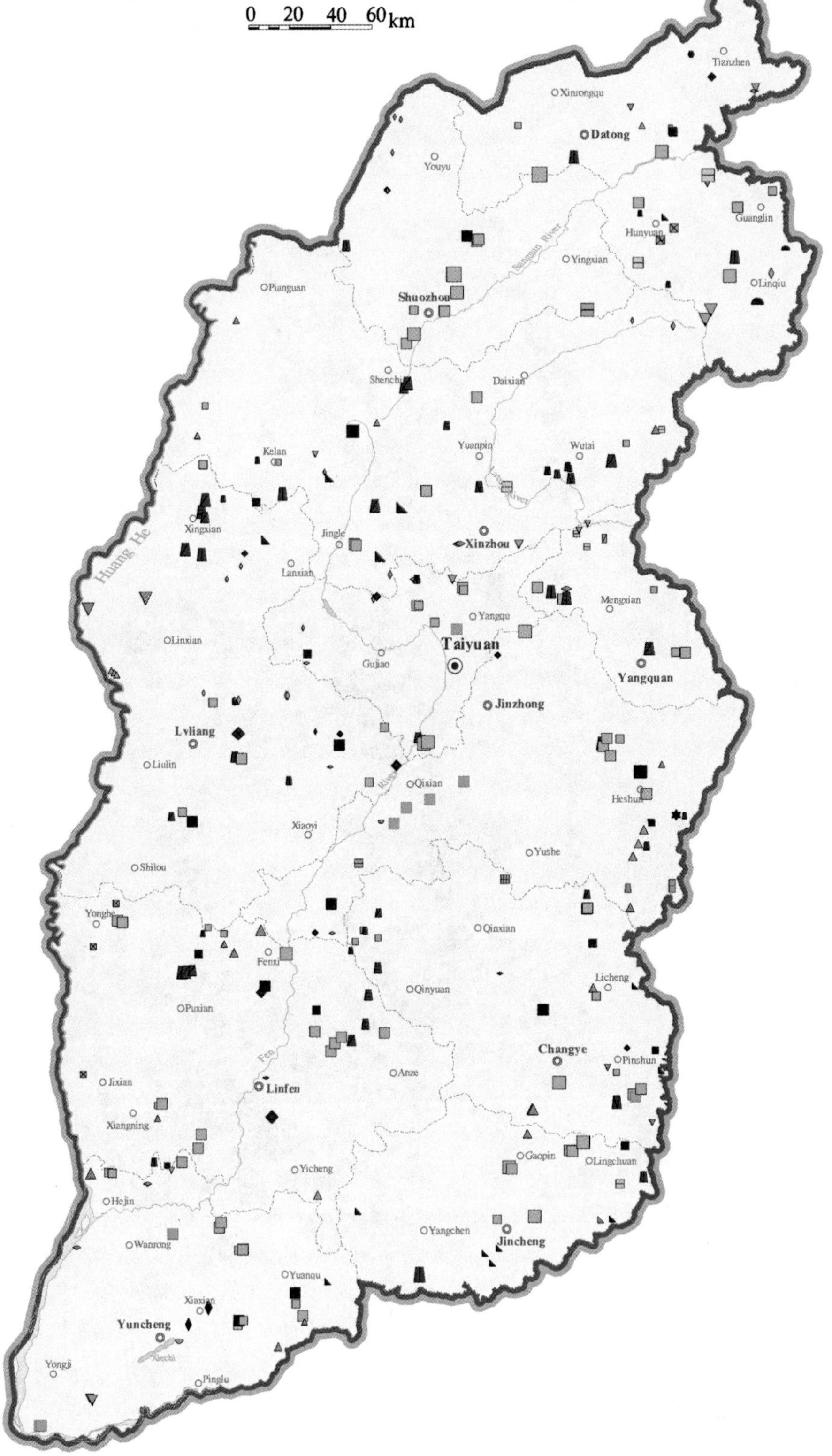

0   20   40   60 km

Tianzhen

Xinrongqu

Datong

Youyu

Guanglin

Hunyuan

Yingxian

Pianguan

Shuozhou

Linqiu

Shenchi

Daixian

Yuanping

Wutai

Kelan

Xingxuan

Jingle

Xinzhou

Lanxian

Mengxian

Linxian

Yangqu

Taiyuan

Gujiao

Yangquan

Lvliang

Jinzhong

Liulin

Qixian

Heshun

Xiaoyi

Yushe

Shilou

Yonghe

Qinxian

Fenxi

Qinyuan

Licheng

Puxian

Changye

Pinshun

Jixian

Linfen

Anze

Xiangning

Gaopin

Lingchuan

Yicheng

Hejin

Wanrong

Yangchen

Jincheng

Yuanqu

Xiaxian

Yuncheng

Yongji

Pinglu

Huang He

Deposit distribution map of Neimongolzizhiqu Region (metallic minerals)

0   100   200   300 km

Deposit distribution map of Neimongolzizhiqu Region
(energy minerals)

0   100   200   300 km

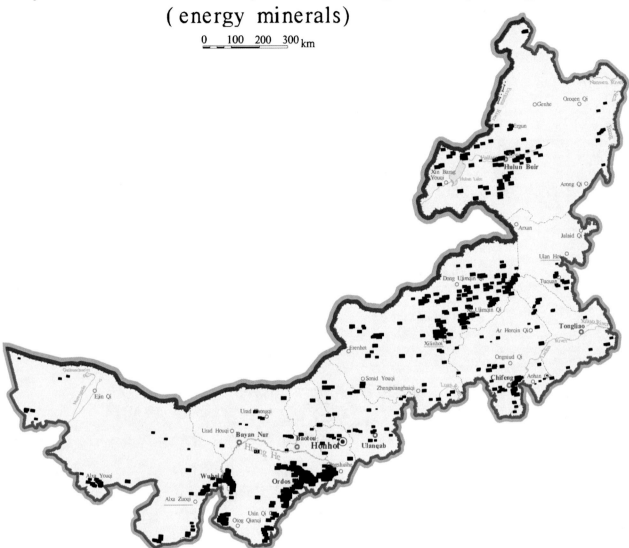

Deposit distribution map of Neimongolzizhiqu Region
(other non-metallic minerals)

0　100　200　300 km

Deposit distribution map of Neimongolzizhiqu Region (building materials)

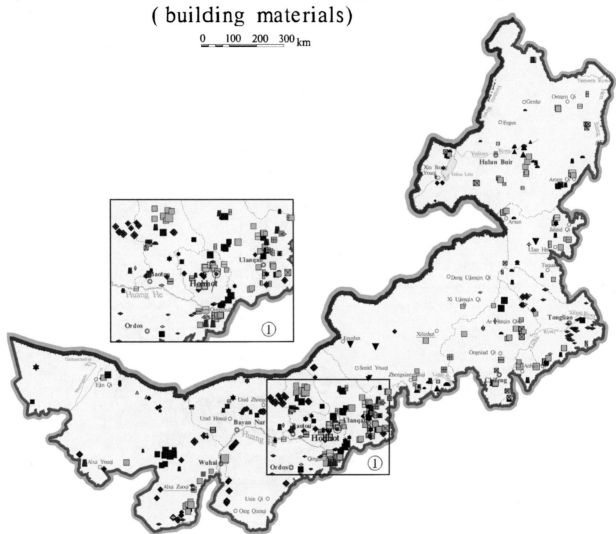

Deposit distribution map of Liaoning Province (metallic minerals)

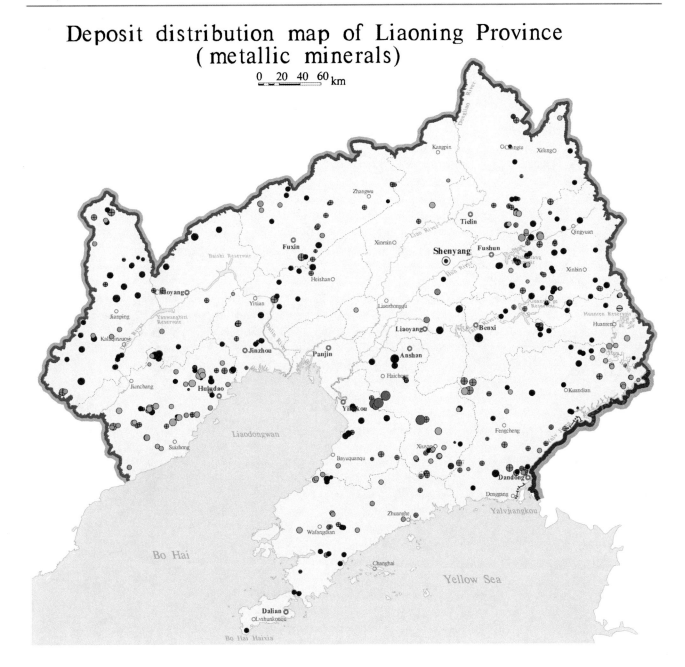

Deposit distribution map of Liaoning Province
(energy minerals)

0   20   40   60 km

Deposit distribution map of Liaoning Province
(non−metallic minerals)

0　20　40　60 km

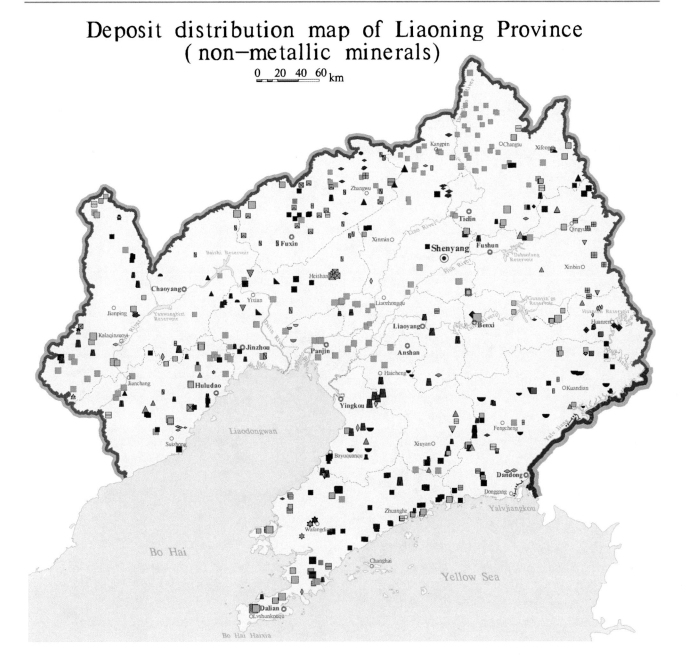

Deposit distribution map of Jilin Province
(metallic minerals)

0 20 40 60 km

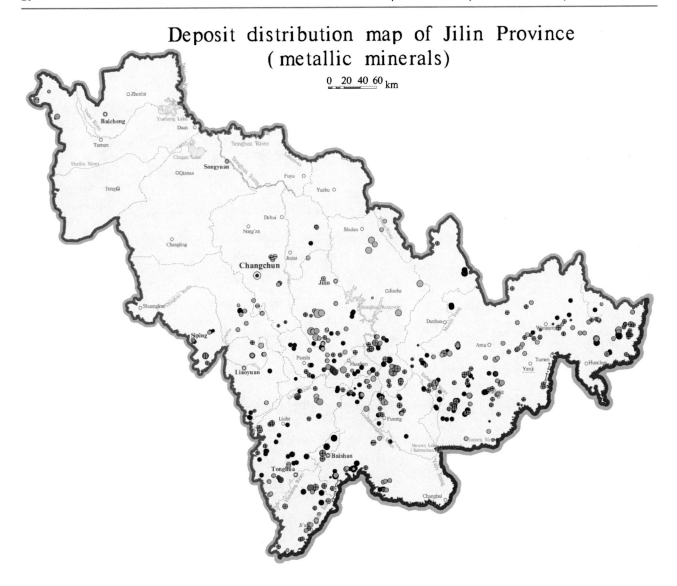

Deposit distribution map of Jilin Province
(energy minerals)

0 20 40 60 km

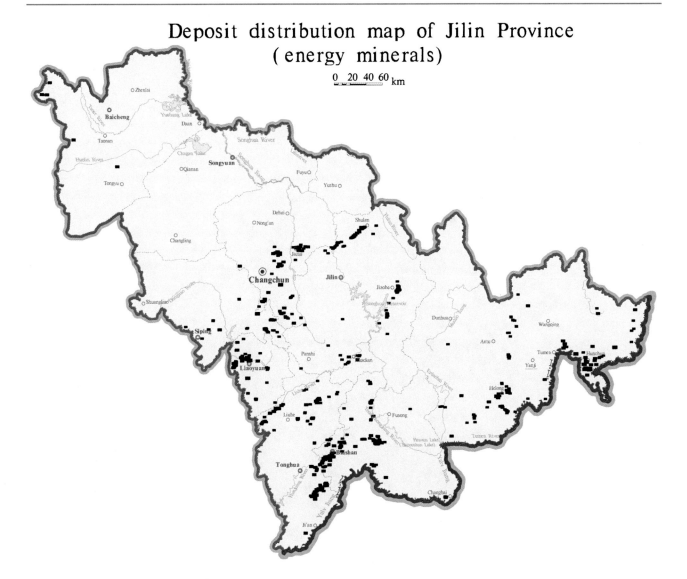

Deposit distribution map of Jilin Province
(non−metallic minerals)

0 20 40 60 km

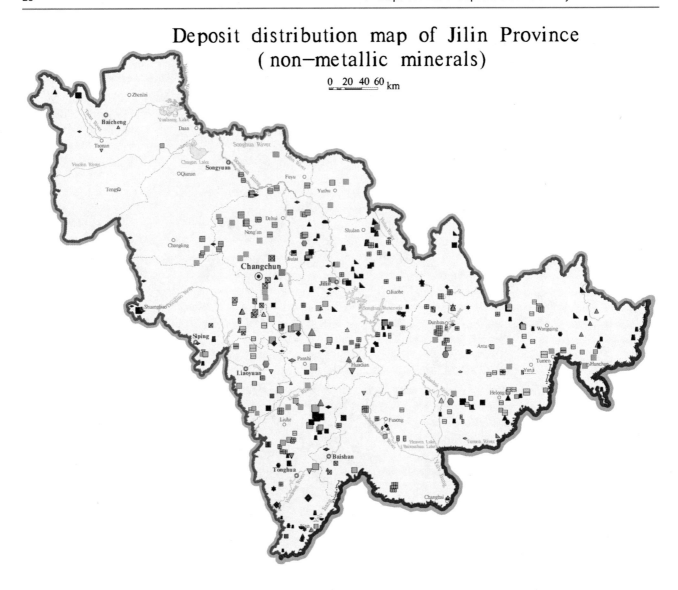

Deposit distribution map of Heilongjiang Province
(metallic minerals)

0   40   80   120 km

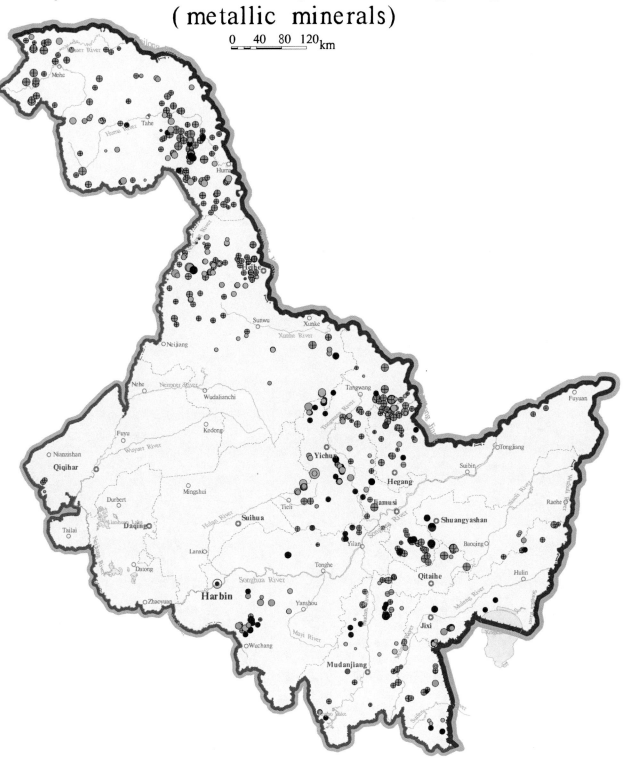

Deposit distribution map of Heilongjiang Province
(energy minerals)

0   40   80   120 km

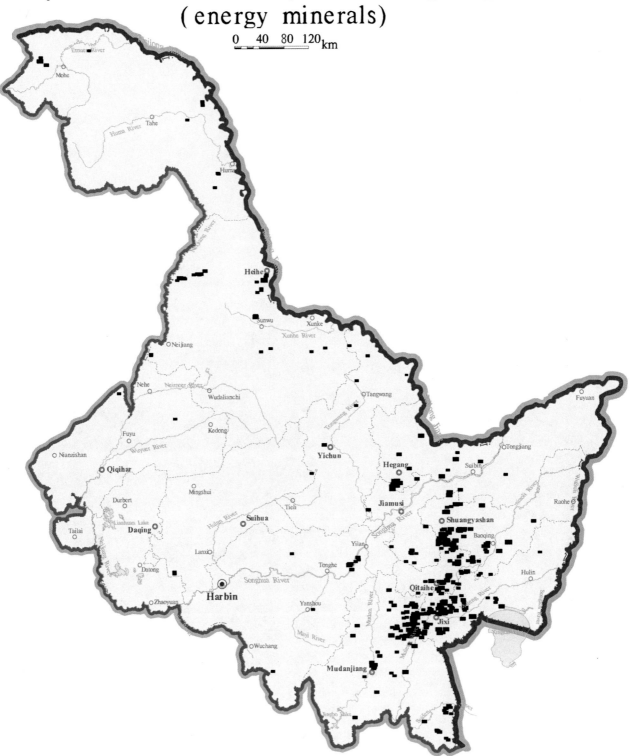

Deposit distribution map of Heilongjiang Province
(non−metallic minerals)

0   40   80 120 km

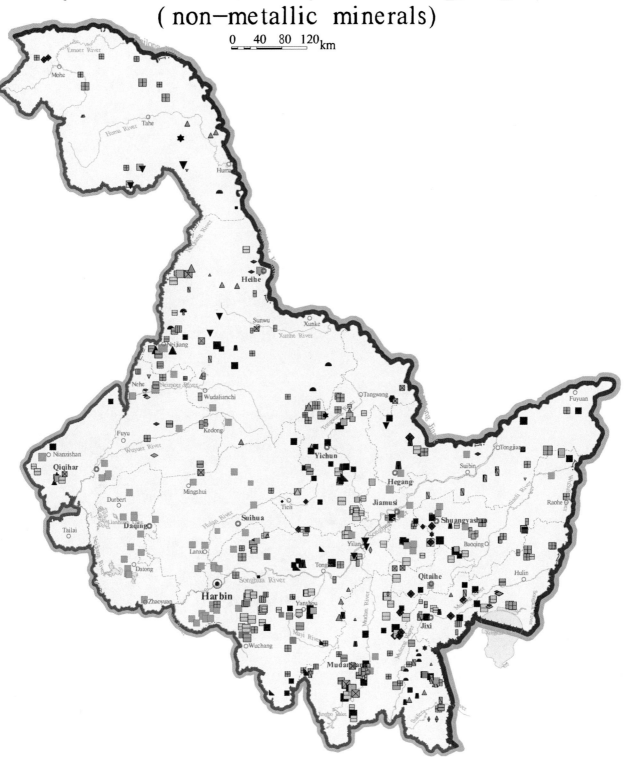

Deposit distribution map of Jiangsu Province
(metallic minerals)

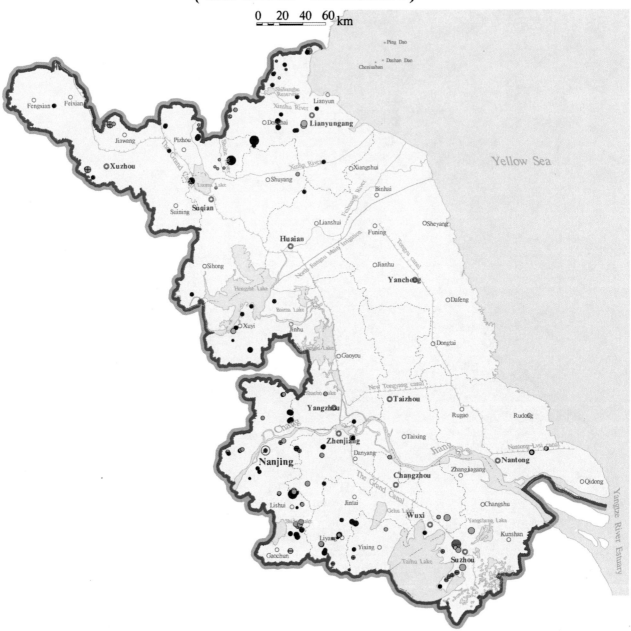

Deposit distribution map of Jiangsu Province
(energy minerals)

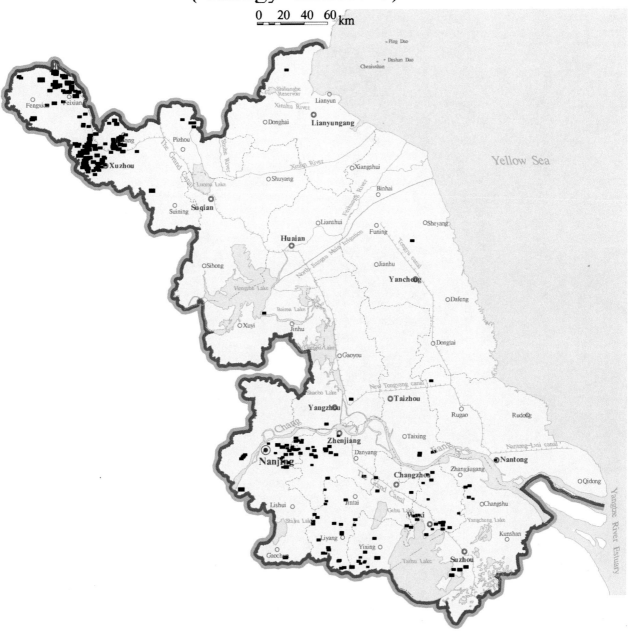

Deposit distribution map of Jiangsu Province
(non−metallic minerals)

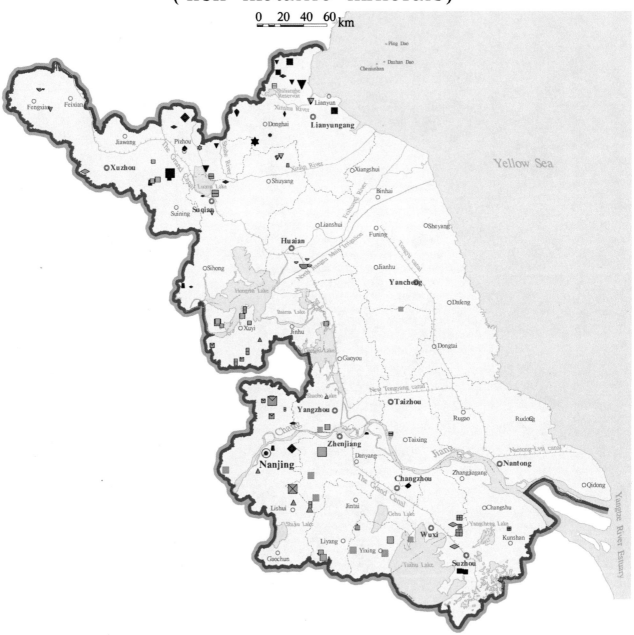

Deposit distribution map of Zhejiang Province
(metallic minerals)

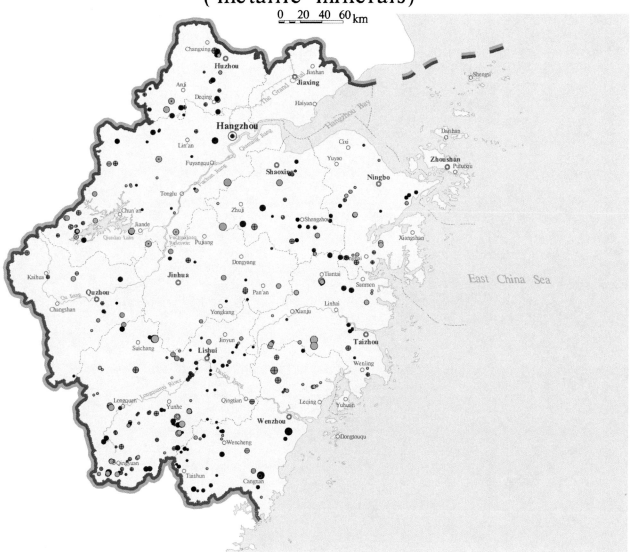

Deposit distribution map of Zhejiang Province
(energy minerals)

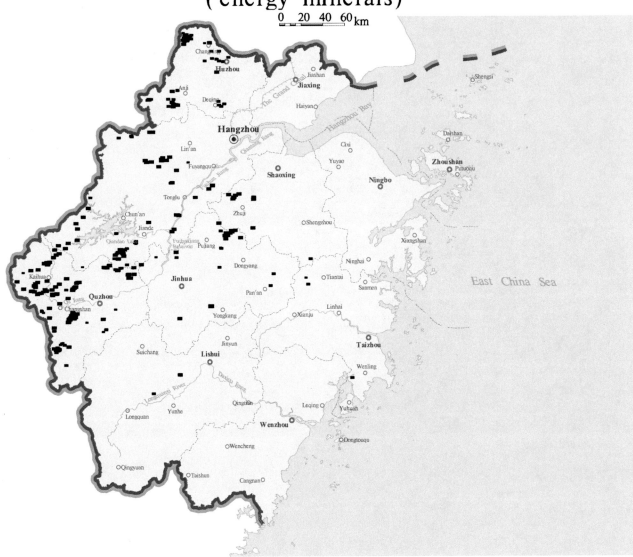

Deposit distribution map of Zhejiang Province
(non−metallic minerals)

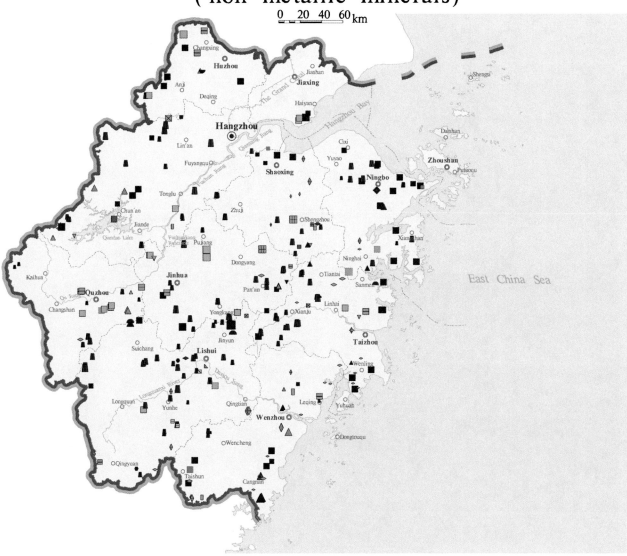

Deposit distribution map of Anhui Province
(metallic minerals)

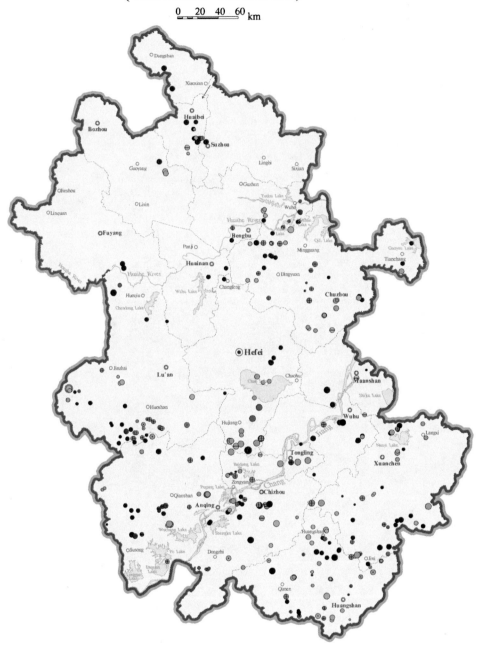

Deposit distribution map of Anhui Province
(energy minerals)

0 20 40 60 km

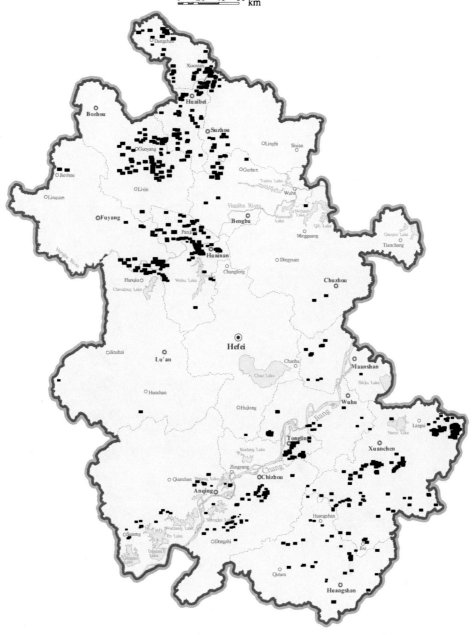

Deposit distribution map of Anhui Province
(non−metallic minerals)

0   20 40 60 km

Deposit distribution map of Fujian Province
(metallic minerals)

0   20   40   60 km

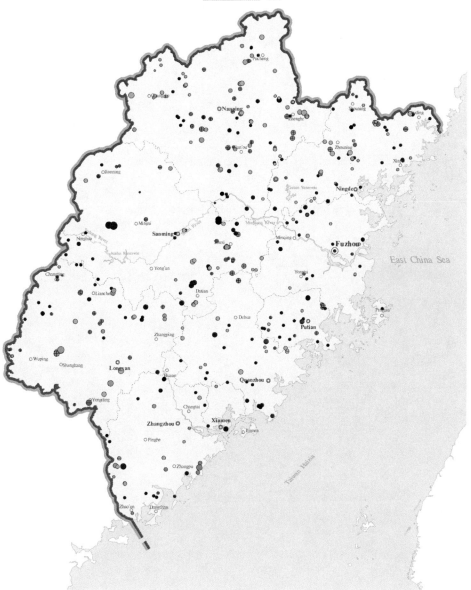

Deposit distribution map of Fujian Province
(energy minerals)

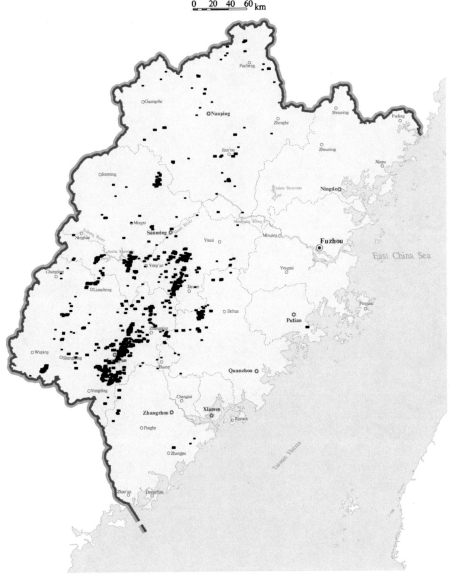

Deposit distribution map of Fujian Province
(non-metallic minerals)

0 20 40 60 km

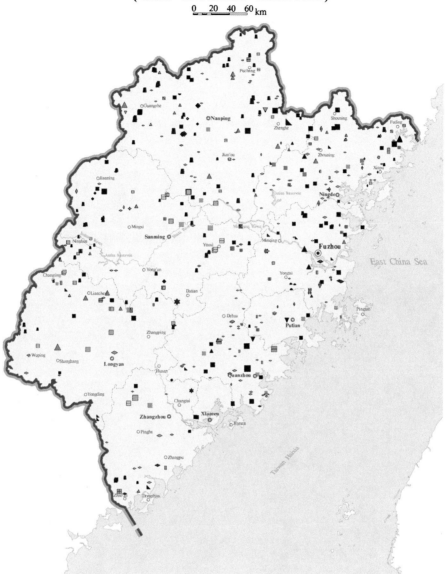

Deposit distribution map of Jiangxi Province
(metallic minerals)

0   20 40 60 km

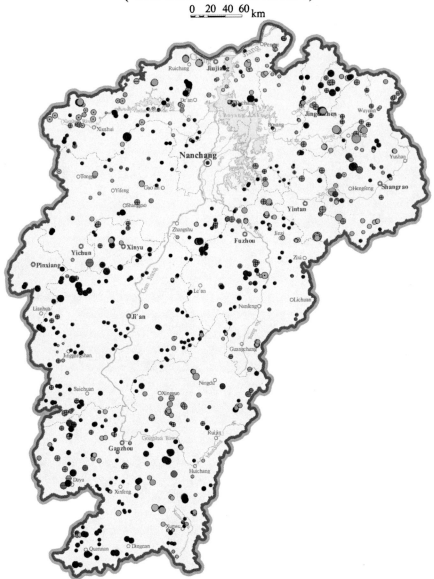

Deposit distribution map of Jiangxi Province
(energy minerals)

0 20 40 60 km

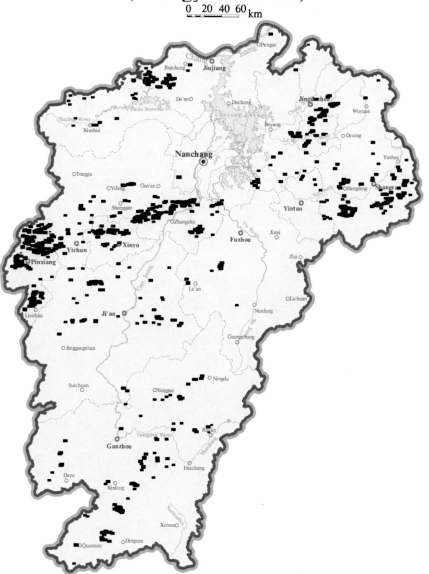

Deposit distribution map of Jiangxi Province
(non—metallic minerals)

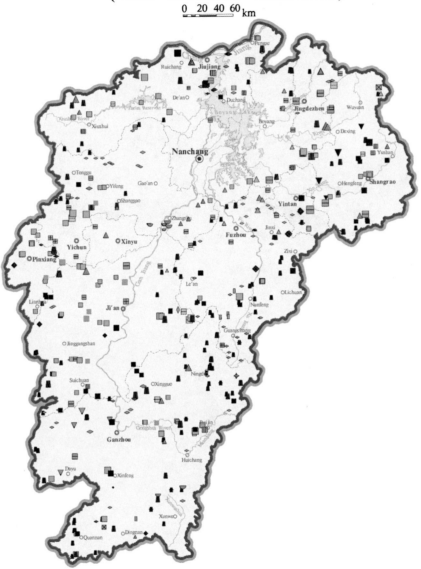

Deposit distribution map of Shandong Province (metallic minerals)

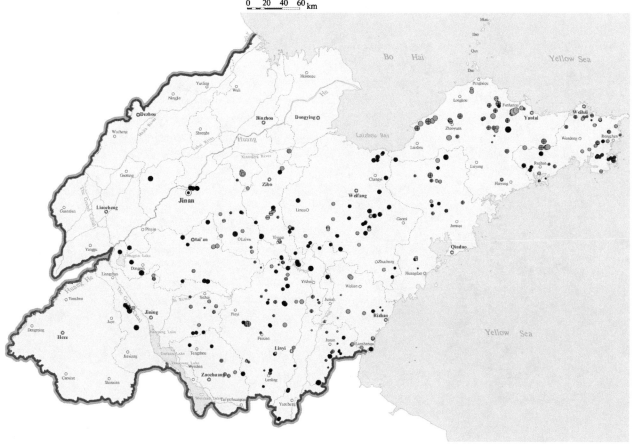

Deposit distribution map of Shandong Province (energy minerals)

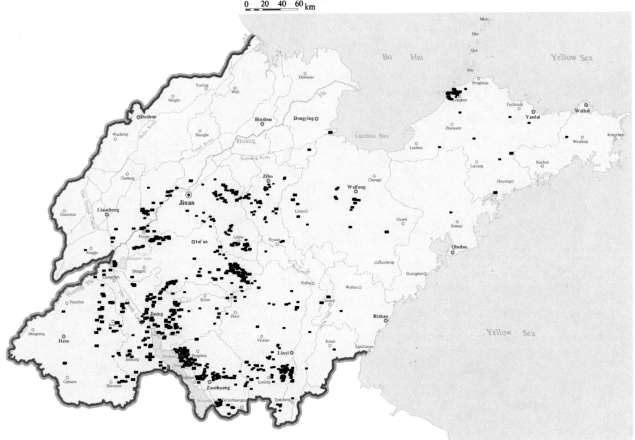

Deposit distribution map of Shandong Province (non-metallic minerals)

0   20   40   60 km

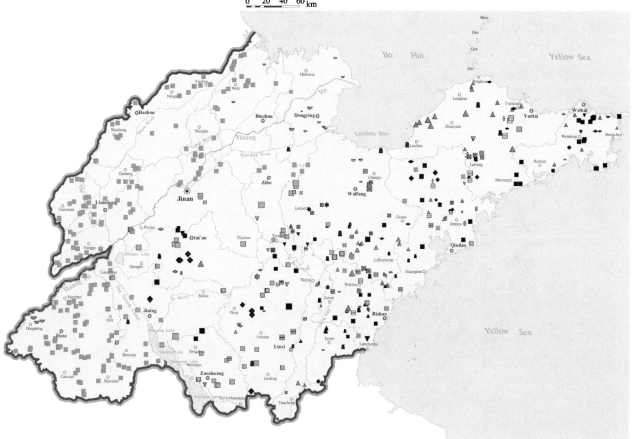

Deposit distribution map of Henan Province
(metallic minerals)

0 20 40 60 km

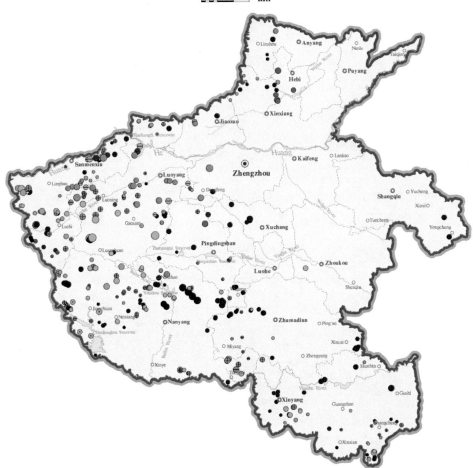

Deposit distribution map of Henan Province
(energy minerals)

0 20 40 60 km

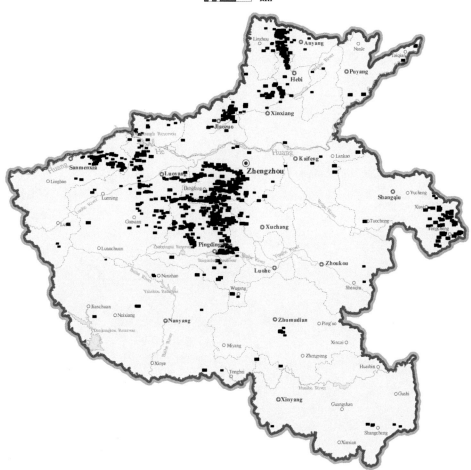

Deposit distribution map of Henan Province
(non-metallic minerals)

0   20 40 60 km

Deposit distribution map of Hubei Province
(metallic minerals)

0   20 40 60 km

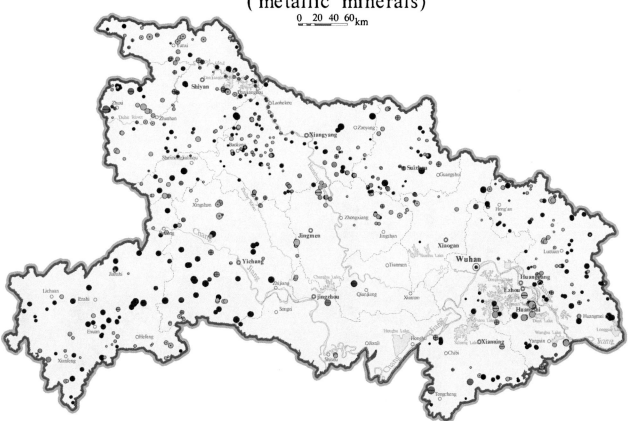

Deposit distribution map of Hubei Province
(energy minerals)

0   20 40 60 km

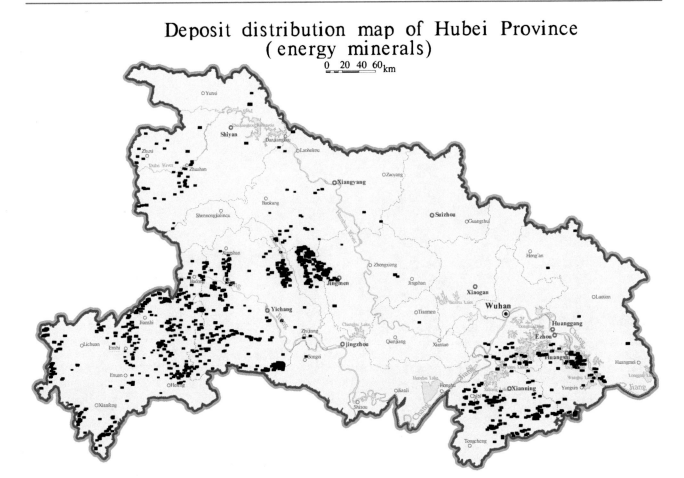

Deposit distribution map of Hubei Province
(non−metallic minerals)

0 20 40 60 km

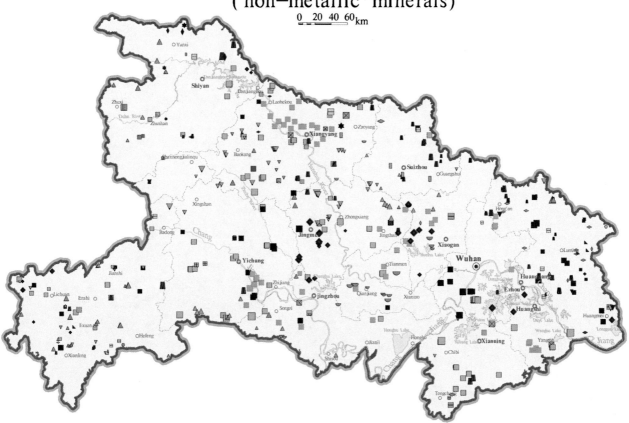

Deposit distribution map of Hunan Province
(metallic minerals)

0 20 40 60 km

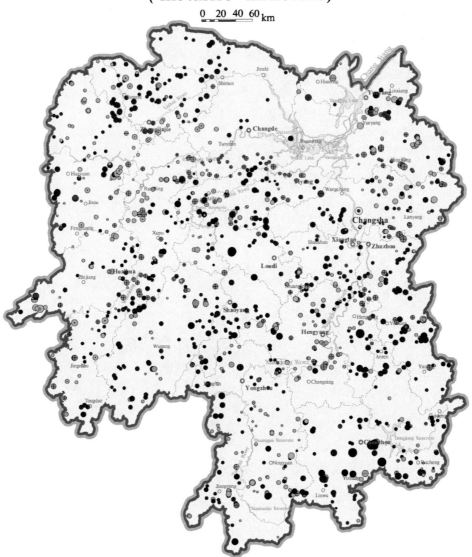

Deposit distribution map of Hunan Province
(energy minerals)

0 20 40 60 km

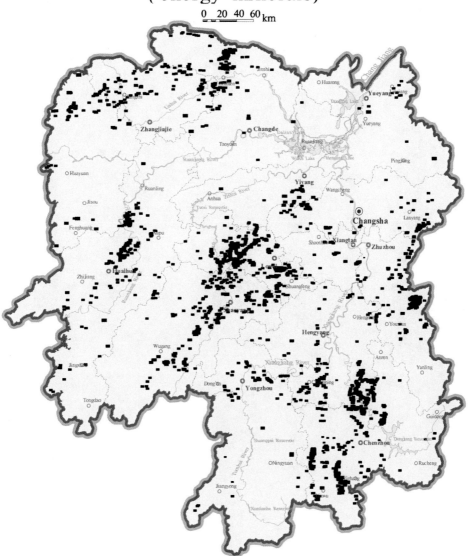

Deposit distribution map of Hunan Province
(non−metallic minerals)

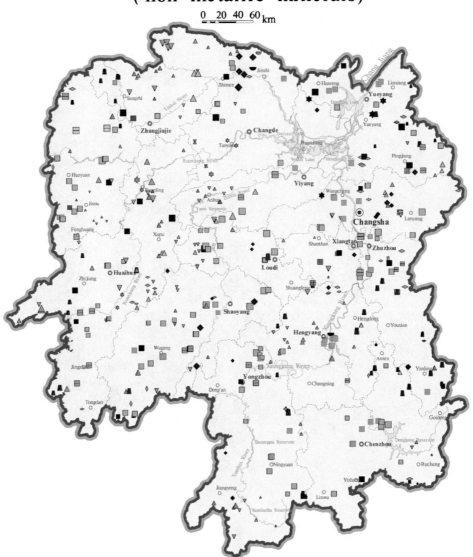

Deposit distribution map of Guangdong Province
(metallic minerals)

0 20 40 60 km

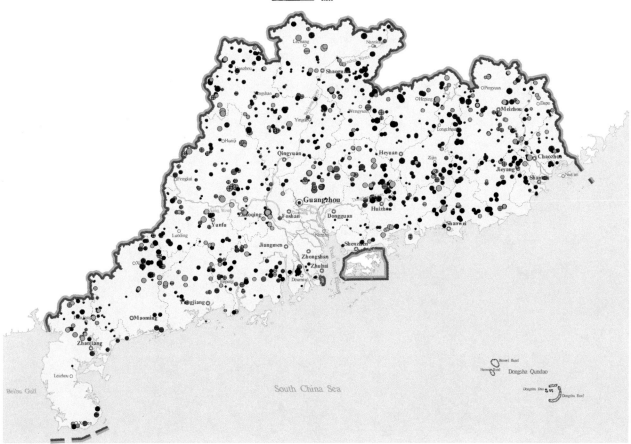

Deposit distribution map of Guangdong Province
(energy minerals)

0 20 40 60 km

Deposit distribution map of Guangdong Province
(non-metallic minerals)

0 20 40 60 km

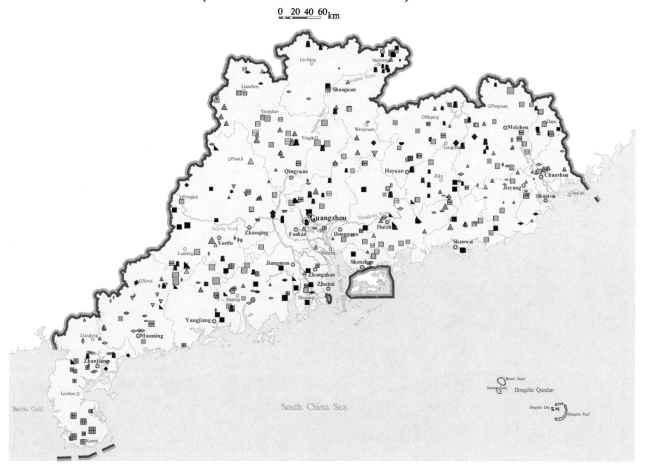

Deposit distribution map of Guangxizhuangzuzizhiqu Region
(metallic minerals)

0   20 40 60 km

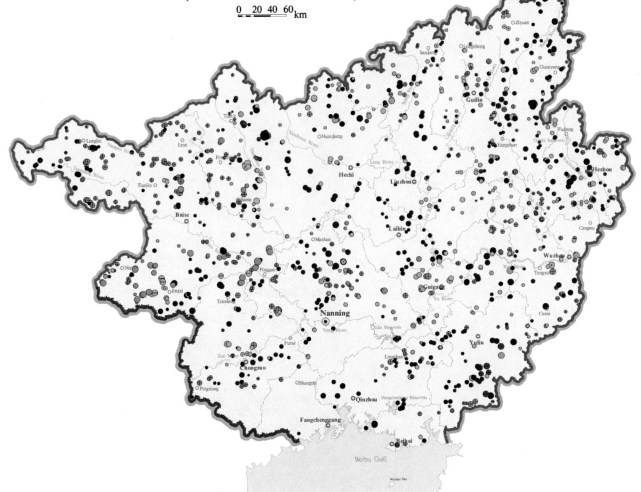

Deposit distribution map of Guangxizhuangzuzizhiqu Region
(energy minerals)

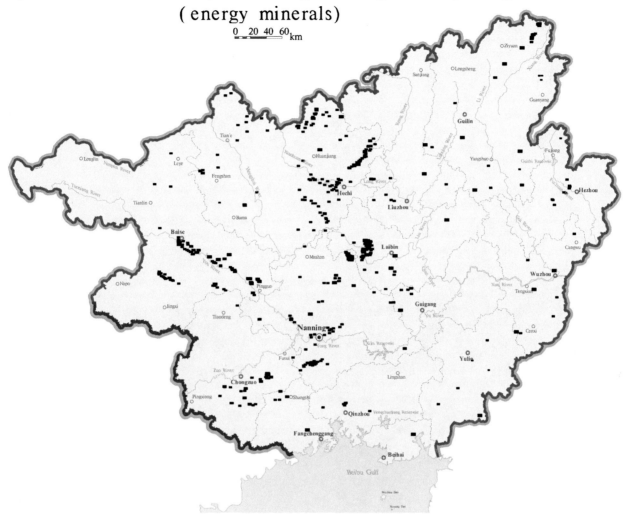

Deposit distribution map of Guangxizhuangzuzizhiqu Region
(non-metallic minerals)

0 20 40 60 km

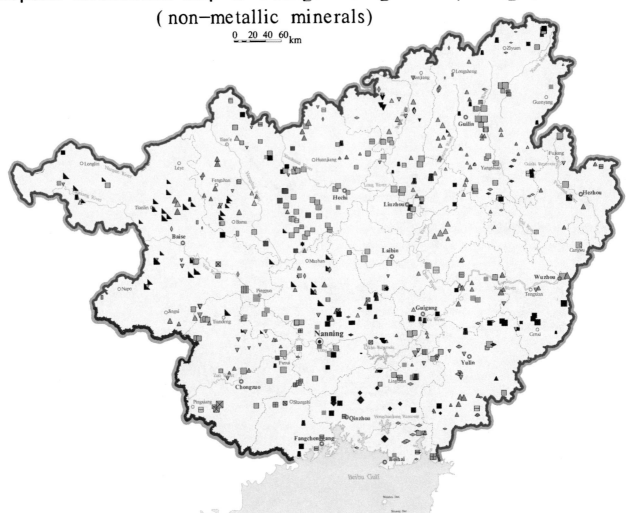

Deposit distribution map of Hainan Province

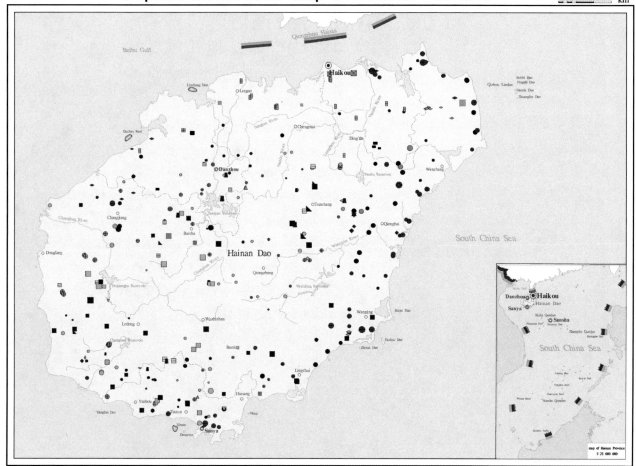

Deposit distribution map of Chongqing Shi, China
(metallic minerals)

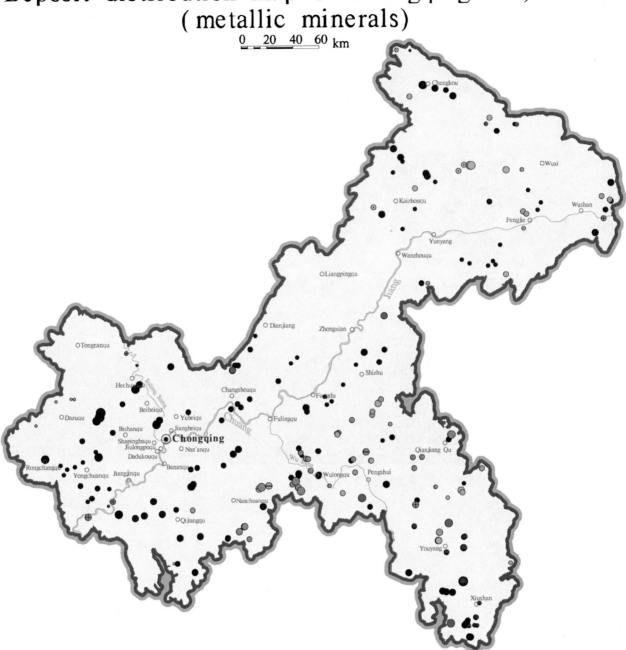

Deposit distribution map of Chongqing Shi, China
(energy minerals)

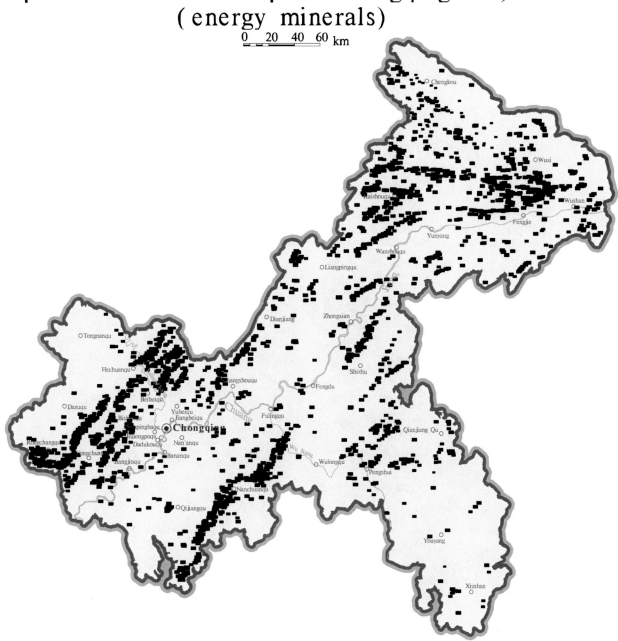

Deposit distribution map of Chongqing Shi, China
(non-metallic minerals)

0   20   40   60 km

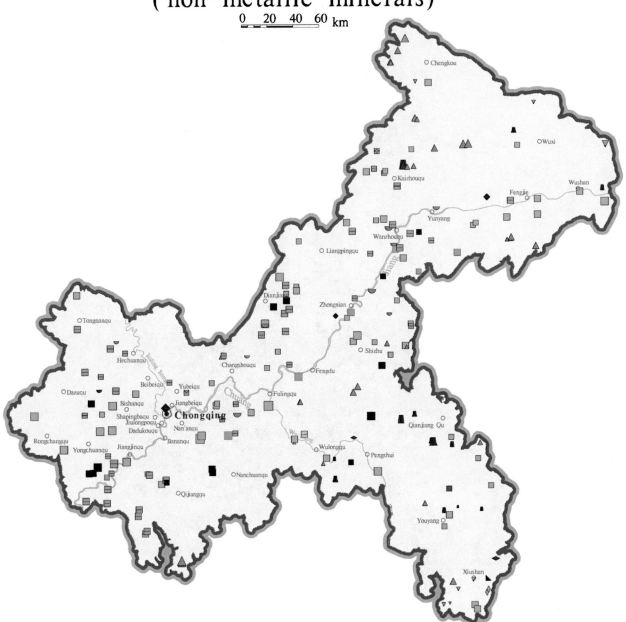

Deposit distribution map of Sichuan Province
(metallic minerals)

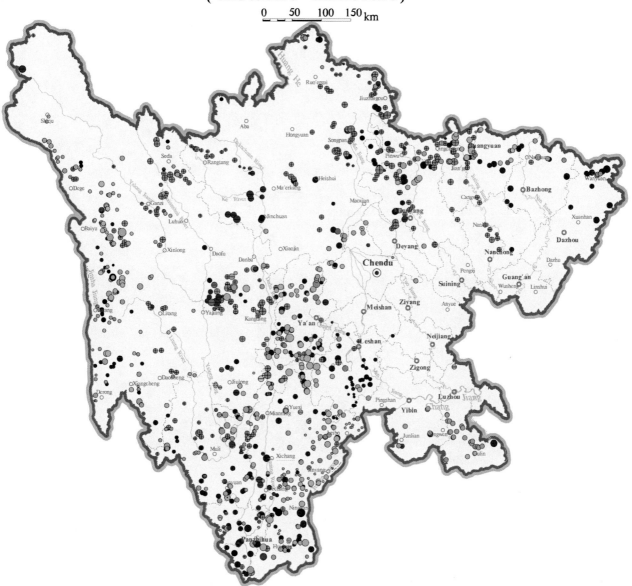

Deposit distribution map of Sichuan Province
(energy minerals)

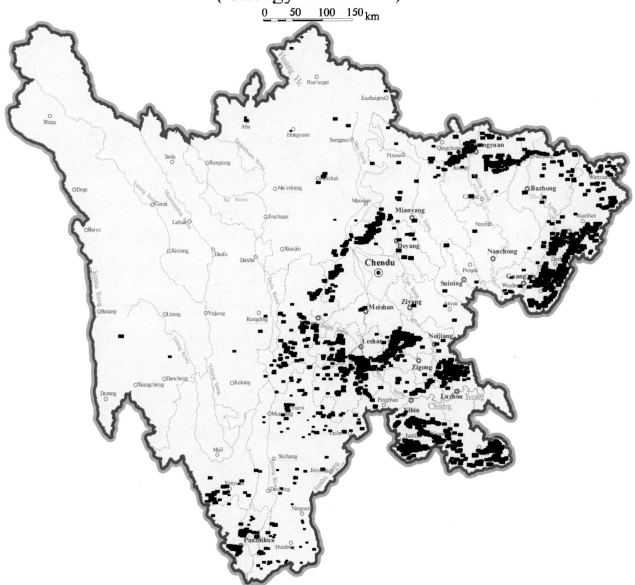

Deposit distribution map of Sichuan Province
(non−metallic minerals)

0   50   100 150 km

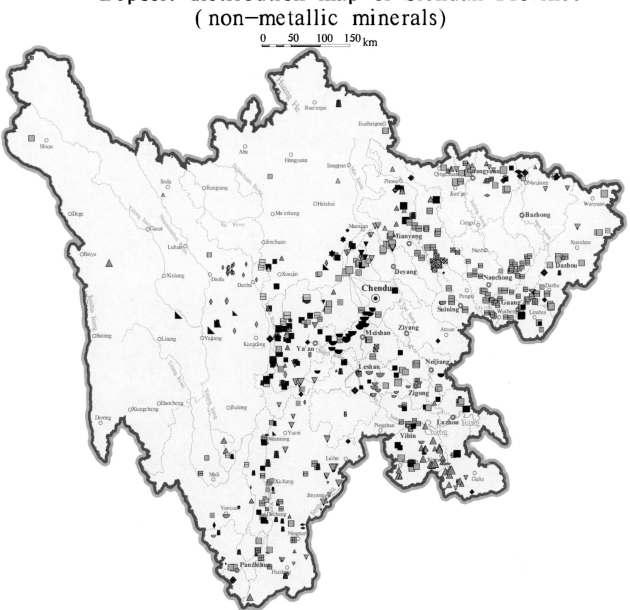

Deposit distribution map of Guizhou Province
(metallic minerals)

0   20   40   60 km

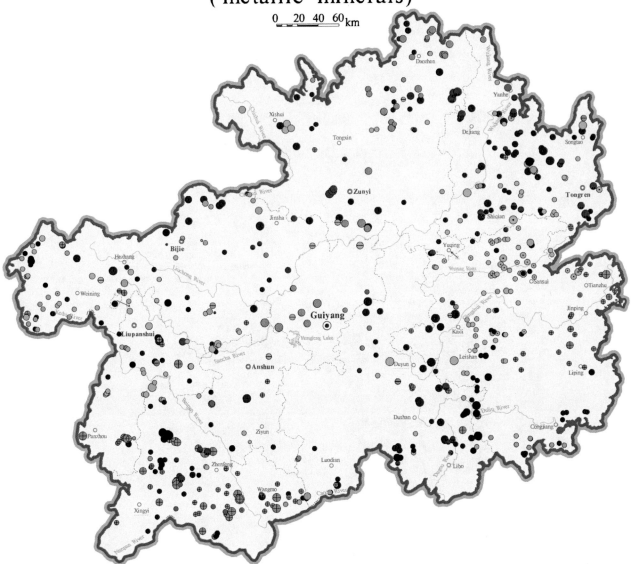

Deposit distribution map of Guizhou Province
(energy minerals)

0   20   40   60 km

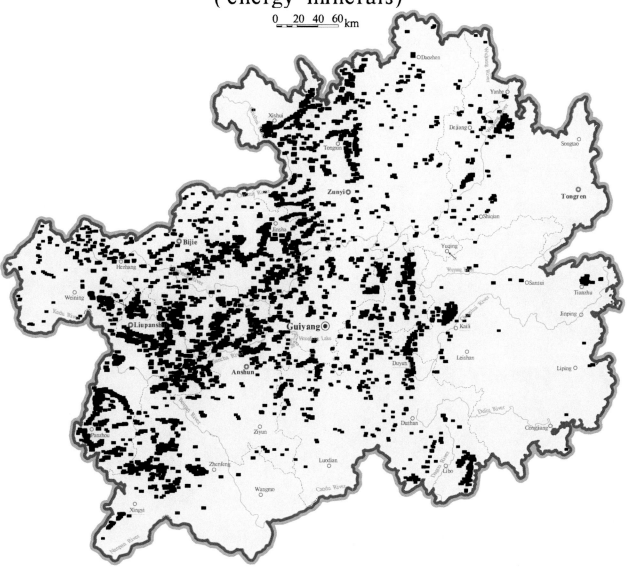

Deposit distribution map of Guizhou Province
(non—metallic minerals)

0   20   40   60 km

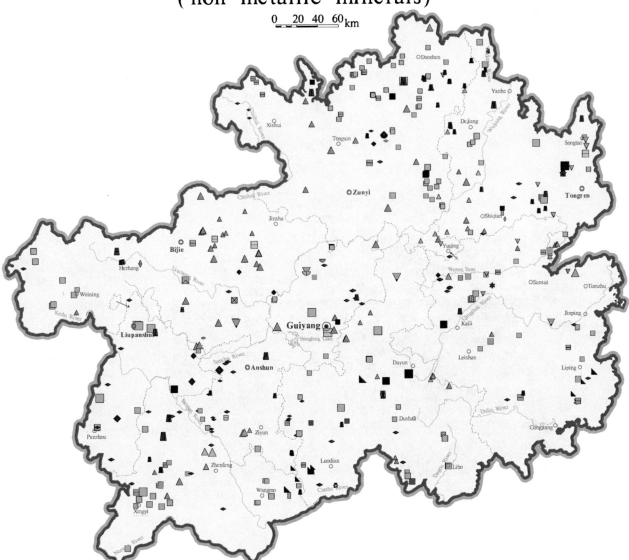

Deposit distribution map of Yunnan Province (metallic minerals)

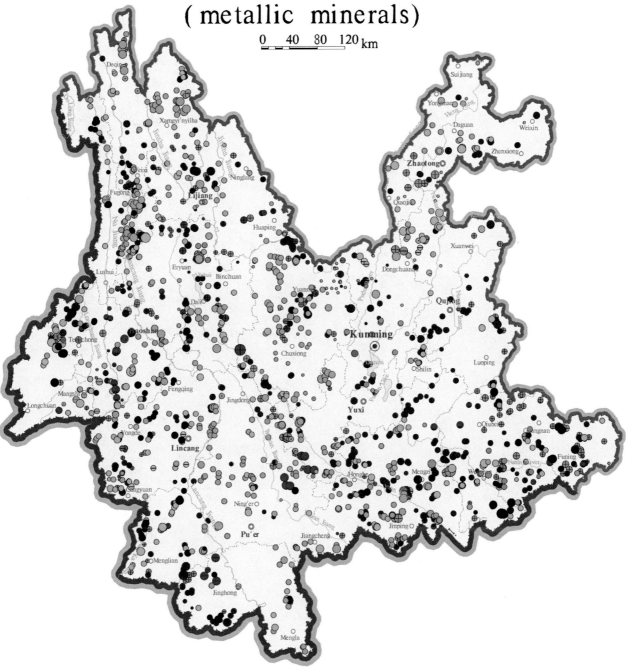

Deposit distribution map of Yunnan Province
(energy minerals)

0   40   80   120 km

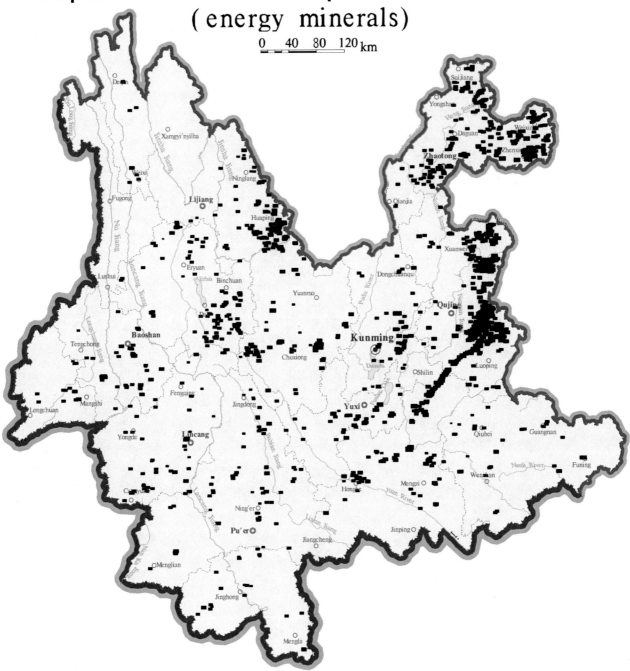

Deposit distribution map of Yunnan Province
(non−metallic minerals)

0   40   80   120 km

Deposit distribution map of Xizangzizhiqu Region
(metallic minerals)

0 50 100 150 km

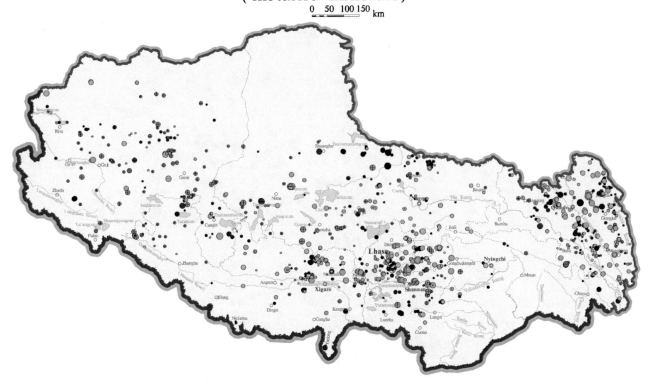

Deposit distribution map of Xizangzizhiqu Region
(non-metallic minerals and energy minerals)

0 50 100 150 km

Deposit distribution map of Shaanxi Province
(metallic minerals)

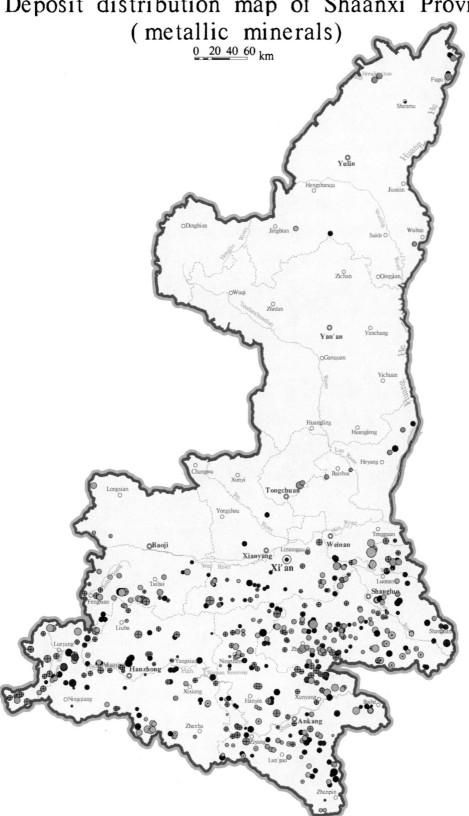

Deposit distribution map of Shaanxi Province
(energy minerals)

0 20 40 60 km

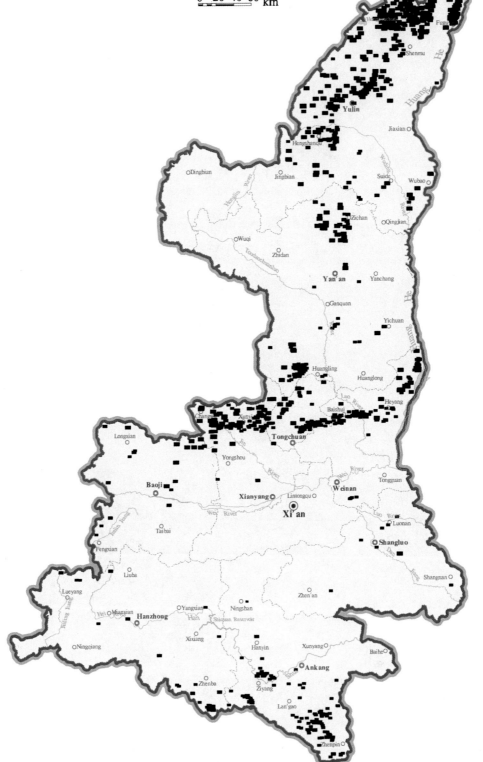

Deposit distribution map of Shaanxi Province
(non-metallic minerals)

0 20 40 60 km

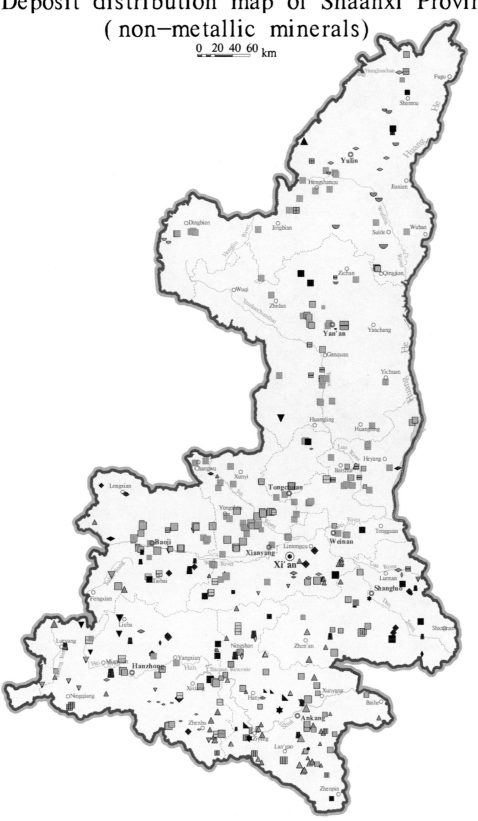

Deposit distribution map of Gansu Province
(metallic minerals)

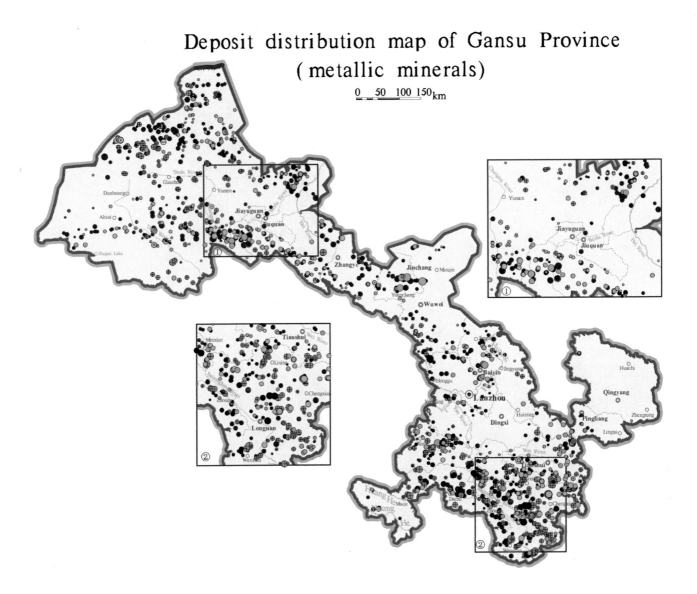

Deposit distribution map of Gansu Province
(energy minerals)

0   50   100   150 km

Deposit distribution map of Gansu Province
(other non−metallic minerals)

0   50   100 150 km

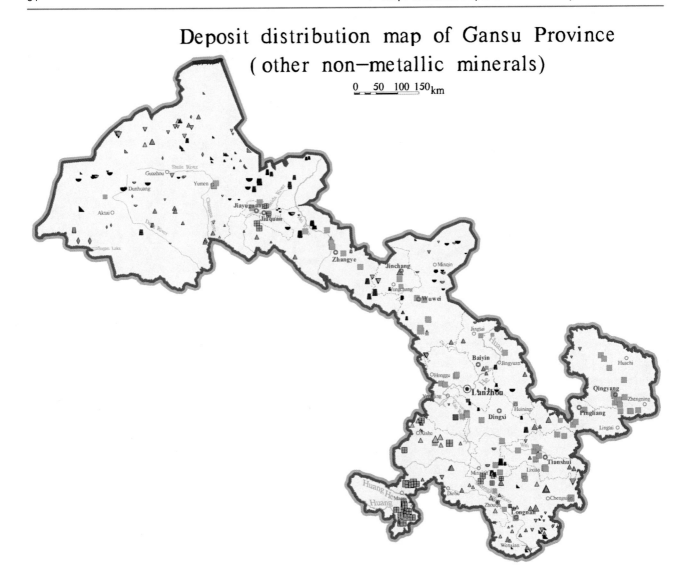

Deposit distribution map of Gansu Province
(building materials)

0   50   100 150km

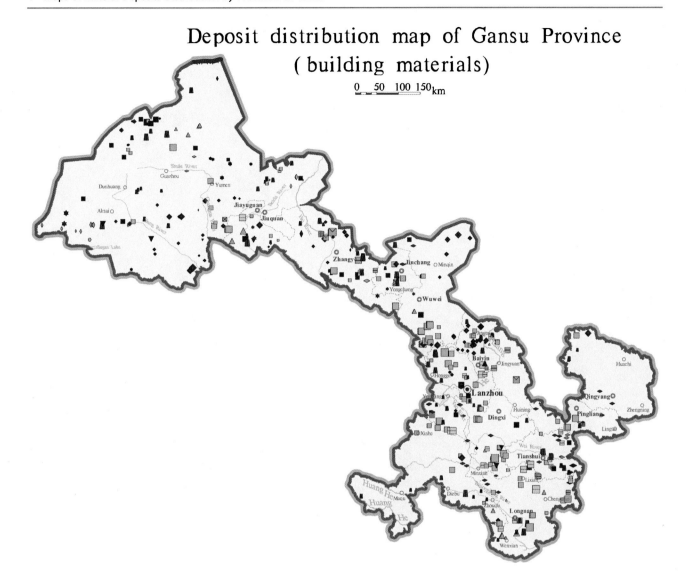

Deposit distribution map of Qinghai Province
(metallic minerals)

0   40   80   120 km

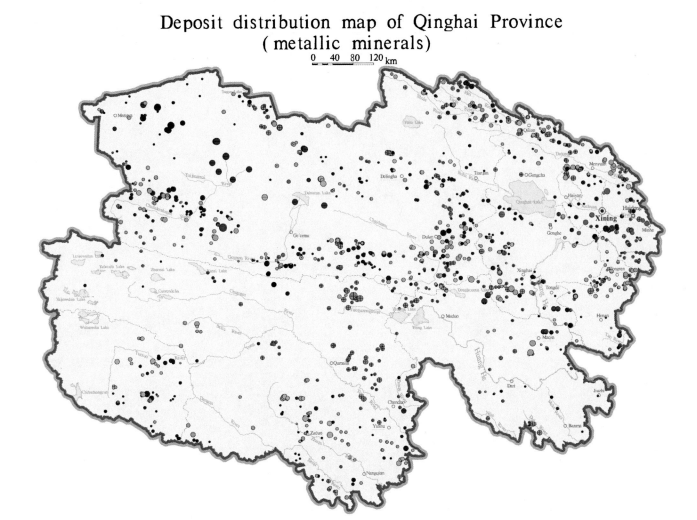

Deposit distribution map of Qinghai Province
(energy minerals)

0   40   80   120 km

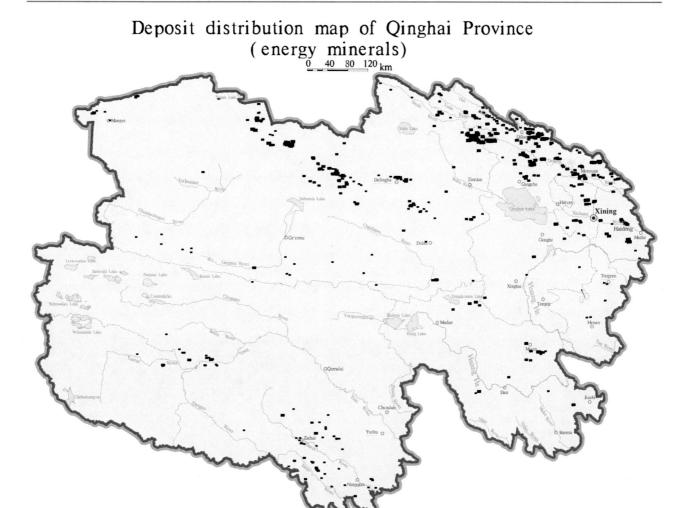

Deposit distribution map of Qinghai Province
(non-metallic minerals)

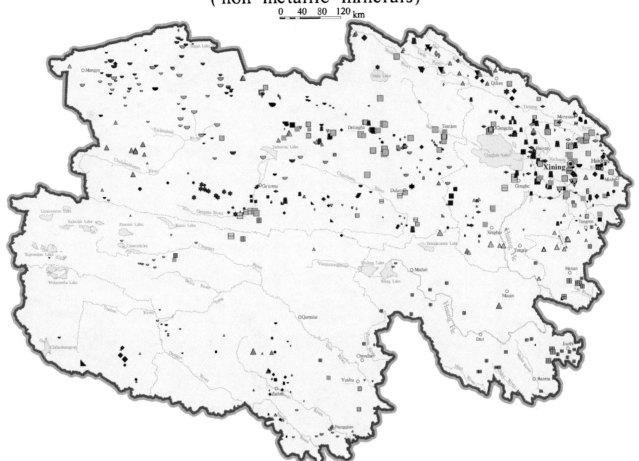

Deposit distribution map of Ningxiahuizuzizhiqu Region

0 10 20 30 km

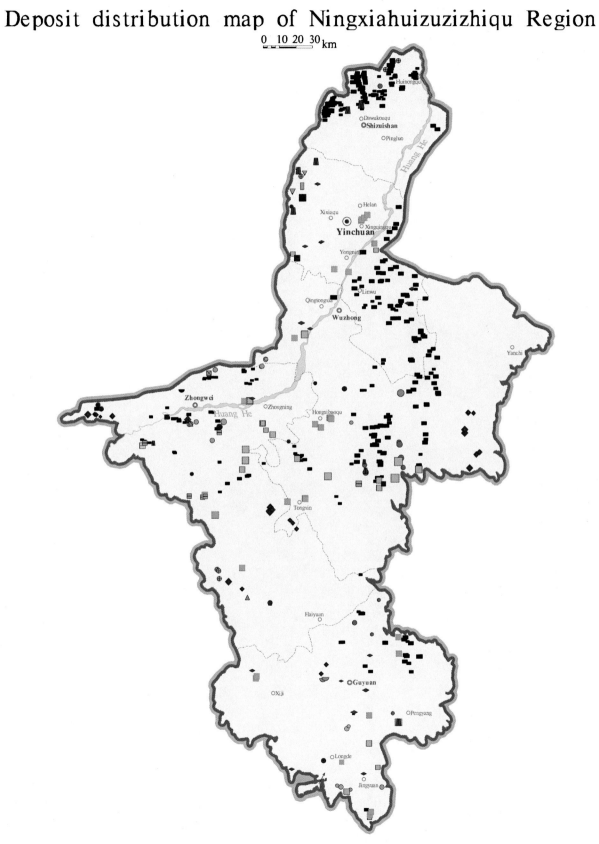

Deposit distribution map of Xinjianguygurzizhiqu Region
(metallic minerals)

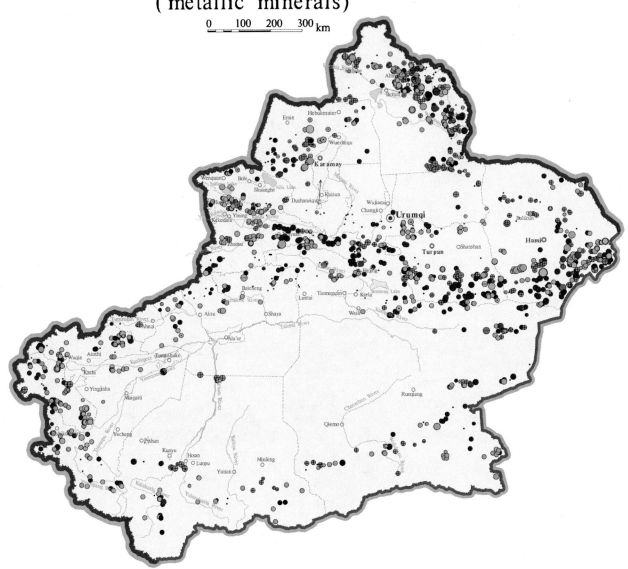

Deposit distribution map of Xinjianguygurzizhiqu Region
(energy minerals)

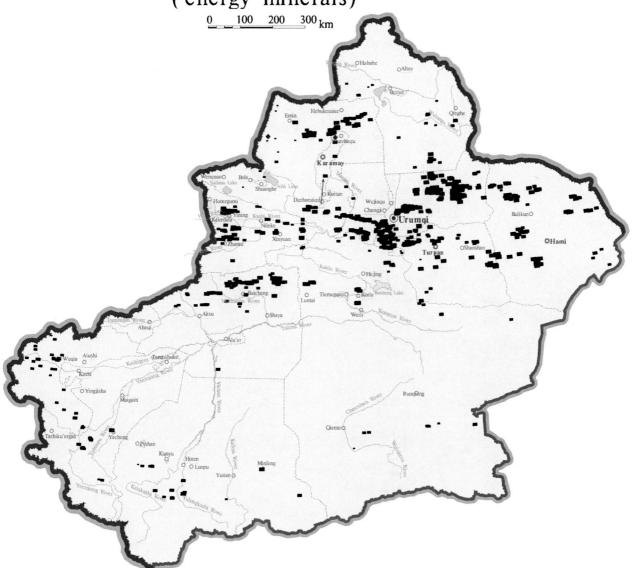

Deposit distribution map of Xinjianguygurzizhiqu Region
(other non-metallic minerals)

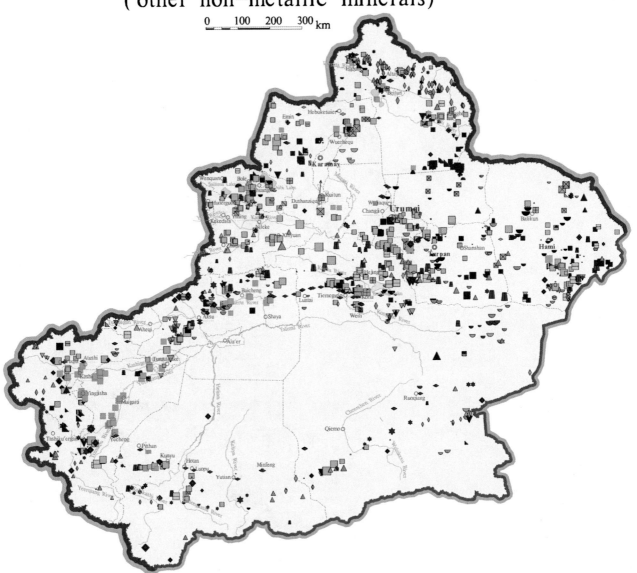

Deposit distribution map of Xinjianguygurzizhiqu Region (building materials)

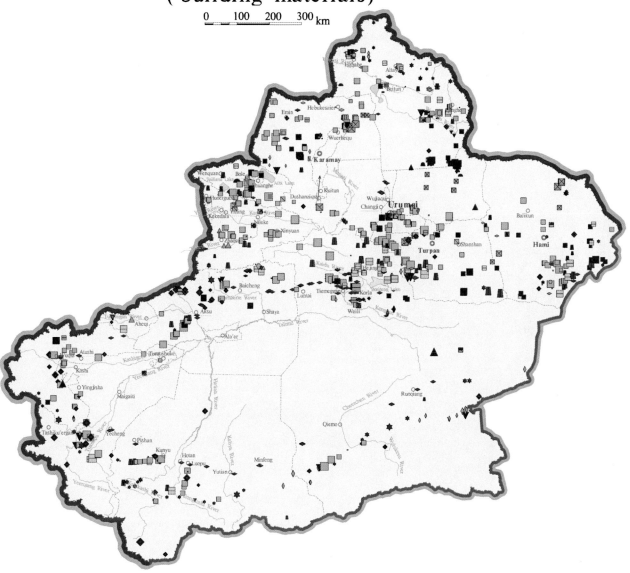

Printed in the United States
by Baker & Taylor Publisher Services